TRANSFRONTIER MOVEMENTS OF HAZARDOUS WASTES

LEGAL AND INSTITUTIONAL ASPECTS

ORGANISATION FOR ECONOMIC CO-OPERATION AND DEVELOPMENT

Pursuant to article 1 of the Convention signed in Paris on 14th December, 1960, and which came into force on 30th September, 1961, the Organisation for Economic Co-operation and Development (OECD) shall promote policies designed:

- to achieve the highest sustainable economic growth and employment and a rising standard of living in Member countries, while maintaining financial stability, and thus to contribute to the development of the world economy;
- to contribute to sound economic expansion in Member as well as non-member countries in the process of economic development; and
- to contribute to the expansion of world trade on a multilateral, non-discriminatory basis in accordance with international obligations.

The Signatories of the Convention on the OECD are Austria, Belgium, Canada, Denmark, France, the Federal Republic of Germany, Greece, Iceland, Ireland, Italy, Luxembourg, the Netherlands, Norway, Portugal, Spain, Sweden, Switzerland, Turkey, the United Kingdom and the United States. The following countries acceded subsequently to this Convention (the dates are those on which the instruments of accession were deposited): Japan (28th April, 1964), Finland (28th January, 1969), Australia (7th June, 1971) and New Zealand (29th May, 1973).

The Socialist Federal Republic of Yugoslavia takes part in certain work of the OECD (agreement of 28th October, 1961).

Publié en français sous le titre:

MOUVEMENTS TRANSFRONTIÈRES
DE DÉCHETS DANGEREUX

The Seminar on Legal and Institutional Aspects of Transfrontier Movements of Hazardous Waste was organised in Paris by the OECD on 12th-14th June 1984 with financial support from the Swiss Government. This meeting was attended by over 80 experts from Governments, professional sectors, academic circles and international organisations, all acting in a personal capacity.

This Seminar, following the adoption by the OECD Council of the Decision and Recommendation on Transfrontier Movements of Hazardous Waste on 1st February 1984, was intended to provide background information necessary for implementation by Member countries of the OECD Decision and Recommendation, as well as to indicate areas where further international cooperation on legal and institutional aspects may be necessary.

It was held in the framework of the OECD Environment Committee's work on hazardous waste and benefitted from the help of the OECD Waste Management Policy Group. This report is published under the responsibility of the Secretary-General and does not necessarily reflect the views of the OECD or its Member governments.

85-8469

TABLE OF CONTENTS

POLICY ISSUES CONCERNING TRANSFRONTIER MOVEMENTS OF HAZARDOUS WASTE

J. W. MacNeill
Director, Environment Directorate
OECD, Paris

TRANSFRONTIER MOVEMENTS: A FOCUS FOR CONCERN

A great deal of public and government attention was focussed on the saga of 41 drums of wastes contaminated with dioxin from the Seveso accident, which were temporarily lost after having crossed the frontier from Italy into France. Happily, these drums were found and their contents are now in the process of being safely disposed. Unhappily, in the year since the Seveso affair took place, a great quantity of hazardous wastes have crossed national frontiers on their way to anonymity rather than to appropriate and safe disposition.

In fact, each hour, about twelve loads of hazardous wastes will cross a national frontier somewhere in OECD Europe. During 1983, approximately 100 000 such border crossings comprising 2.2 million tonnes of waste, occurred in Europe, while more than 5 000 occurred in North America. In other words, in the OECD, a cargo of hazardous wastes crosses a national frontier more than once every five minutes, 24 hours a day, 365 days per year. Most waste is transported by road in 22-24 tonne lots; some travels by rail. We at OECD are not aware of transport of significant amounts of waste by inland waterway or by air.

Many of these cargoes move across frontiers for reasons which are both sensible and legal. But certain crossings occur because controls are inadequate. If this trend were to proceed in an unanticipated and uncontrolled fashion, the Seveso affair might be but a drop in a potentially very large and menacing bucket.

There are strong indications that the amount of waste which crosses national frontiers is increasing and that this upward trend is likely to continue. In Europe between 1982 and 1983, wastes transported for disposal in another country -- i.e. in a country other than the country of origin -- virtually doubled. About 700 000 tonnes of waste fell into this category. This increase may be attributed partly to the recent availability of relatively low-cost, legal land-based disposal facilities in certain non-OECD countries.

One reason for the increase in transfrontier movements is that a number of countries are strengthening internal control over hazardous wastes. In many cases, wastes deemed to be highly hazardous will be required to be specially treated or destroyed by incineration. These regulations often result in the imposition of certain additional, or incremental costs, to the generators of such wastes. For example, a specific waste which was formerly disposed in the land may be required by regulation to be disposed by incineration. For the generator, who previously might have landfilled the waste at the site of origin, this means that the waste must be catalogued, packaged, transported to the incinerator and finally destroyed. This can cost up to an additional $75 per tonne on average for highly hazardous waste. Despite these additional costs, there is a definite trend on the part of OECD Member governments to anticipate possible damages from hazardous wastes and to attempt to prevent these damages from occurring by careful internal control of the wastes.

But internal control is only part of the story. In these times of increasing flows of goods across national frontiers, and decreasing customs formalities, especially within trading blocs, countries have understandably grown concerned about the movement of hazardous wastes across national frontiers. The Decision and Recommendation adopted by the OECD Council Concerning Transfrontier Movements of Hazardous Waste in early 1984 provides the foundation for this Seminar. The European Economic Community also is currently preparing a Directive meant to provide a basis for governing transfrontier flows of hazardous wastes between EEC members. Recent legislation in the United States requires close scrutiny of such exports and imports by a Federal agency.

To respond to these concerns, the OECD has been accelerating its efforts to lay the basis for an international system to make the necessary control over transfrontier movements a practical reality.

THE REASONS FOR TRANSFRONTIER MOVEMENTS

Lest I give the impression that the OECD is against any transfrontier movements of hazardous wastes whatsoever, let me dwell for a moment on some of the reasons which may underlie such movements.

Individual firms, acting in their economic interest, will naturally seek to minimise incremental costs arising from the need to comply with governmental control of hazardous wastes. If costs, including all administrative and transport expenses, are lower when legal treatment, storage and/or disposal of hazardous wastes can be accomplished in a country other than that of origin, then there will be motivation for transfrontier shipment of hazardous wastes.

Why should the costs be lower elsewhere? Treatment, storage and/or disposal costs, including transport expenses, may be lower in the importing country because better and more recent technology is available there. They may be lower because the importing country's facility can take advantage of economies of scale, serving a greater or a more concentrated industrial area. Or the reason may be quite clear if one looks at the map. An appropriate treatment, storage and/or disposal facility in the destination country may be nearer to the generation site than any such facility in the home country.

There may be additional valid reasons for transfrontier movement of hazardous wastes, e.g.:

-- no firm in the exporting country has the technology to treat the wastes to the standard required in that country; or

-- a multinational corporation which has provided for an appropriate facility at one of its sites may elect to transport all suitable wastes from its sites in other countries to that facility.

In certain cases, less expensive legal disposal costs for a unit weight of hazardous waste may be available because controls are less stringent in the importing country as compared to the exporting country. So long as this situation exists, certain generators will be strongly motivated to utilise the less expensive option. There may also be a certain temptation to "lose" waste in a country which has no strong internal control system meant to assure that wastes arrive at a licensed facility.

There is another factor favouring globetrotting of hazardous wastes. A simple analysis of shipping costs suggests that transport of certain highly hazardous wastes, (e.g. those banned from sea dumping by international conventions), via ship from industrialised nations to certain less developed countries could be very profitable in economic terms for all participating parties. Hence, the occurrence of a "North-to-South" movement of highly hazardous wastes should be considered a real possibility.

As noted above, given existing costs for hazardous waste disposal, an individual generator can save $75 per tonne on average if legal land based disposal is available as opposed to more expensive disposal methods such as incineration. Given costs to transport one tonne of waste a distance of one kilometer, a journey by road or rail of 800 km can be contemplated by individual generators in order to reduce disposal costs for wastes which would otherwise require disposal methods such as incineration or physico-chemical treatment.

Given the location of the population centres of OECD Europe, European Members of OECD most likely to be concerned with traffic in such highly hazardous wastes include: Austria, Belgium, Denmark, France, Germany, Italy, Luxembourg, Netherlands and Switzerland. In addition, given the cost differentials for various disposal methodologies, hazardous waste which crosses frontiers destined for disposal in another country is likely to be waste considered highly hazardous, and thus requiring incineration or physico-chemical treatment in the generator country, as well as being restricted from legal sea dumping by one or more international conventions.

What do the generators here in Europe gain in aggregate savings from shipping 700 000 tonnes of their highly hazardous wastes about the countryside? At absolute maximum, the saving is 700 000 times $75 or $52.5 million.

And how does this compare with the total cost of disposal of potentially hazardous waste in Europe? It is barely four per cent! (Europe spent roughly $1.3 billion to dispose of hazardous wastes in 1983.)

I believe that this situation is something of the tail wagging the dog. It is, however, rather a potent tail in terms of its capacity for damaging our environment. Clearly, with more than 100 000 international yearly transports of potentially hazardous waste moving about Europe, the possibility for "loss" of considerable amounts of material -- in 23 tonne lots -- is very large. The point is that "saving" the generators about $52 million per year may well cost many times that amount in possible remedial action costs. Experience has shown that the cost of cleaning up damaged areas caused by a unit weight of hazardous waste can be well over 100 times the cost of appropriate disposal of the material. In other words, "loss" of 0.2 per cent of the material being transported could cost more than $52 million to clean up. And the burden would generally be on the public not the polluter.

SOME POLICY ISSUES IN TRANSFRONTIER MOVEMENTS OF HAZARDOUS WASTE

The policy issues involved in transfrontier movement, can be examined by taking as a starting point the Principles which were contained in the OECD Recommendation.

Apart from the principal reason outlined above, why do we need international cooperation in this field? Are not existing arrangements, for example customs requirements, adequate to cope with the problem? There are several reasons why this is not the case. Perhaps the most fundamental one is that countries do not define hazardous wastes in the same way -- "hazardous" being inevitably a rather imprecise term. The usual practical method of classification appears to be to list substances contained in the waste which cause it to be classified as hazardous, and such lists have already been developed in a number of countries. But these country lists are not necessarily the same. Consequently, some type of cross-referencing of national lists is necessary in order to alert governments as to a particular type of waste which is subject to controls in their country but to different or even no controls in a possible country of origin or destination, and of course vice versa.

It is clearly impossible for customs authorities to take the responsibility for checking all individual shipments of hazardous waste as they cross frontiers. Not only are the numbers of such shipments very large but these wastes are often very complex, and to verify at the frontier that what is in the truck is really what is specified on the travel documents is an impossible task in practical terms.

Consequently, an international system of notification is called for in the OECD Principles. The system must satisfy three basic requirements -- those of awareness, of identification, and of protection. To meet these goals, the international system must provide the authorities of the countries concerned with adequate and timely information regarding prospective transfrontier movements. It must be designed in such a way as to minimise the possibility of inappropriate transfrontier movement and/or improper disposal of hazardous wastes which may occur because of differences in national policies or control requirements. And an effective monitoring system must not add unduly to the administrative burdens on governments and industry.

10

The design and implementation of such a system raises a number of legal and institutional issues of some complexity. I should like to allude to a few of what I would consider to be the thornier issues.

The legal systems of different Member countries handle liability for damage from international transport of hazardous waste to humans and/or the environment in different ways. Hence, some agreement on methods of procedure may be required when waste from a generator in one country, transported by a firm licensed to do business internationally, is alleged to have created problems in a second country.

Insurance meant to provide financial guarantees against accident or mishap may be extremely expensive or even difficult to obtain in the case of transfrontier hazardous waste transport. The question of reasonable and equitable means to assess the risks involved needs to be settled in order to provide a sounder basis for developing such insurance policies.

An internationally acceptable -- yet uncomplicated and effective -- transport document needs to be developed and tested. The type and quantity of information on such a manifest and how, and by whom, the information will be monitored (and in which languages) needs to be agreed as well.

Before concluding, I should like to return to a couple of issues which I believe to be of considerable importance to the success of our future work in implementing the OECD Council Decision -- even though both issues are "in the wings" of the Decision, so to speak.

The first issue is that of harmonising national systems for managing hazardous wastes. To the extent that we succeed in doing so, the risk of transfrontier movements "for the wrong reasons" will diminish. So our future work programme needs, I think, to give some priority attention to this matter. One idea which could help this process of harmonisation, particularly in Europe, might be the establishment of regional facilities providing appropriate treatment, storage and disposal facilities. Looking at harmonisation in the broader perspective, it is likely that our cooperation in designing an international system will itself exert an influence to bring our national systems into closer harmony. But we need to keep this goal clearly in sight as we proceed with the implementation of the Council Decision.

The second issue concerns those countries, in what is popularly referred to as the "East" and the "South", who are not Members of the OECD, and who would not normally be protected by the international system for controlling transfrontier movements which we are putting into place for the OECD area. It is clearly not the objective of an OECD system to shift the burden for hazardous waste disposal to non-Member countries, and we must make sure that what is not an objective does not become, by accident or bad design, a consequence of the system. So we must, as we construct the OECD system, be outward-looking, not with the intent of imposing any controls on non-Member countries -- which we clearly cannot do -- but with the objective of ensuring that waste generators in OECD countries act in the same responsible manner for exports of their waste to a non-OECD country as they would to one which is bound by our Council Decision. By "outward-looking" of course, I also mean that we need to cooperate closely with other international bodies active in this field, whose membership includes regional groupings outside the OECD, and the more global community. Only in this way can we move towards a truly effective international system for controlling hazardous waste.

TOWARDS BETTER COORDINATION OF NATIONAL POLICIES AND INTERNATIONAL COOPERATION

In the OECD International Conference on Environment and Economics, one of the principal topics for discussion is that of Future directions for Environmental Policies. The types of policy available for environmental protection can be broadly characterised as falling into one of two categories -- "React-and-Cure" or "Anticipate-and-Prevent". The latter, if properly designed and administered, is usually more effective and cheaper in the long run, and it is this type of policy which countries are increasingly bringing into being for their internal management of hazardous wastes.

"Anticipate-and-Prevent" policies mean devoting extra care, attention and funds now in order to prevent future problems of the type that happened at Lekkerkerk, Love Canal and about 4 000 other waste disposal sites in OECD countries. But these policies cannot be judged successful if they result in vastly increased quantities of highly toxic wastes being transported with inadequate controls to destinations outside the country of generation so that they can be disposed of at a substantial cost saving to the generator, but at increased risk of considerable long-run cost to society. This would be a case of "Anticipate-and-Prevent" policies increasing the need for later implementation of "React-and-Cure" ones -- albeit in another country -- clearly the worst of both possible worlds.

We see therefore that the implementation of "Anticipate-and-Prevent" policies internally must be accompanied by effective controls over movements of the wastes. To achieve this, international cooperation is an imperative. It must accompany, and be accompanied by, strong, effective and reasonably harmonised national laws and regulations meant to provide adequate control over hazardous wastes. Without one, the other cannot succeed.

DECISION AND RECOMMENDATION ON TRANSFRONTIER MOVEMENTS OF HAZARDOUS WASTE
C(83)180(Final)

(Adopted by the OECD Council at its 598th Meeting
on 1st February, 1984)

The Council,

Having regard to Articles 5(a) and 5(b) of the Convention on the Organisation for Economic Co-operation and Development of 14th December, 1960;

Having regard to the Recommendation of the Council of 28th September, 1976 on a Comprehensive Waste Management Policy [C(76)155(Final)];

Bearing in mind that the Governments of the OECD Member countries have recognised "the responsibility they share to safeguard and improve the quality of the environment, both nationally and in a global context" and have declared that "the protection and progressive improvement of the quality of the environment is a major objective of the OECD Member countries" (Declaration on Environmental Policy, 1974);

Considering that a number of OECD Member countries generate substantial amounts of hazardous waste and that a significant proportion of such waste is subject to transfrontier movements;

Considering that efficient and environmentally sound management of hazardous waste may justify some transfrontier movements of such waste in order to make use of appropriate disposal facilities in other countries;

Considering that the generator of a hazardous waste has responsibilities to ensure that the disposal of its waste is carried out in a manner consistent with the protection of the environment, whatever the place of disposal;

Considering that countries have the sovereign right to manage hazardous waste within their jurisdiction pursuant to their own environmental policies and legislation, taking account of the rules of international law;

Considering the need for concerted action among Member countries to protect man and his environment against pollution which may arise in connection with hazardous waste management;

On the proposal of the Environment Committee:

I.   DECIDES that Member countries shall control the transfrontier movements of hazardous waste and, for this purpose, shall ensure that the competent authorities of the countries concerned are provided with adequate and timely information concerning such movements.

II.   RECOMMENDS that, to implement this Decision, countries apply the principles concerning transfrontier movements of hazardous waste set out below.

III.   INSTRUCTS the Environment Committee, having regard to work of other international organisations, to elaborate a programme of activities to develop further the principles set out below and facilitate their implementation, and to explore what additional international action may be necessary concerning transfrontier movements of hazardous waste.

IV.   INSTRUCTS the Environment Committee to review periodically action taken by Member countries in pursuance of this Decision and Recommendation.

*   *
*

PRINCIPLES CONCERNING TRANSFRONTIER MOVEMENTS OF HAZARDOUS WASTE

The following principles are designed to facilitate the development of harmonized policies concerning transfrontier movements of hazardous waste. They do not prejudice the implementation of more favourable measures for the protection of the environment that are now in force or that may be adopted; neither do they prejudice the application of any international agreement dealing with the free trade of goods or services or the transport of dangerous goods.

Definitions of terms used in these Principles are given in the Appendix.

General principles

1.   Countries should ensure that hazardous waste situated within the limits of their jurisdiction is managed in such a way as to protect man and the environment. For this purpose, countries should promote the establishment of appropriate disposal installations and should adopt all necessary measures to enable their authorities to control the activities related to generation, transport and disposal of hazardous waste, and to ensure compliance with the laws and regulations in force.

2.   In respect of the management of hazardous waste that is subject to transfrontier movements countries should require that:

   a) the entities concerned abstain from participation in transfrontier movements which do not comply with the laws and regulations applicable in the countries concerned;

b) the entities involved in transport or disposal be authorized for this purpose.

3.   Furthermore, with regard to any specific transfrontier movement of hazardous waste, countries should require that the generator of the waste should:

a) take all practicable steps to ensure that the transport and disposal of its waste be undertaken in accordance with the laws and regulations applicable in the countries concerned;

b) in particular obtain assurances that all entities concerned with the transfrontier movement or the disposal of its waste have the necessary authorisations to perform their activities in accordance with the laws and regulations applicable in the countries concerned;

c) reassume responsibility for the proper management of its waste, including if necessary the re-importation of such waste, if arrangements for safe disposal cannot be completed.

4.   Countries should apply their laws and regulations on control of hazardous waste movements as stringently in the case of waste intended for export as in the case of waste managed domestically.

International Pre-Notification and Cooperation

5.   Countries should co-operate in the control, from the place of generation to the place of disposal, of all hazardous waste that is subject to transfrontier movements.

5.1   For this purpose, and given the Decision, countries should take the measures necessary to ensure that the entities within their jurisdiction provide, directly or indirectly, the authorities of the exporting, importing and transit countries with adequate and timely information.

5.2   This information should specify the origin, nature, composition, and quantities of waste intended to be exported, the conditions of carriage, the nature of environmental risks involved, the type of disposal and the identity of all entities concerned with the transfrontier movement or the disposal of the waste.

6.   Exporting countries should take the measures necessary to ensure that a request from an importing or transit country for relevant information elicits a constructive and diligent response.

7.   Countries should adopt the measures necessary to enable their authorities to object to or, if necessary, prohibit the entrance of a consignment of hazardous waste into their territory, for either disposal or transit, if the information provided is insufficient or inaccurate or the arrangements made for transport or disposal are not in conformity with their legislation.

8.   Countries should take all practicable steps to ensure that a projected transfrontier movement of hazardous waste is not initiated if one of the countries concerned has decided in conformity with its legislation to oppose the import or transit of the waste and has so informed the entities or authorities concerned in the exporting country.

9.   When an importing or transit country opposes in conformity with its legislation a transfrontier movement into its territory and the waste has already left the exporting country, the latter should not oppose reimport of the waste.

Appendix

Definitions

For the purposes of the above principles:

a) Waste means any material considered as waste or legally defined as waste in the country where it is situated or through or to which it is conveyed;

b) Hazardous waste means any waste other than radioactive waste considered as hazardous or legally defined as hazardous in the country where it is situated or through or to which it is conveyed, because of the potential risk to man or the environment likely to result from an accident or from improper transport or disposal;

c) Transfrontier movement of hazardous waste means any shipment of waste from one country to another, where the waste is considered as being hazardous waste in at least one of the countries concerned. Hazardous waste arising from the normal operation of ships, including slops and residues, shall not be considered a transfrontier movement covered by this Decision and Recommendation;

d) Exporting country means any country from which a transfrontier movement of hazardous waste is initiated or is envisaged;

e) Importing country means any country to which a transfrontier movement of hazardous waste takes place or is envisaged for purpose of disposal (treatment, landfill, storage, dumping or incineration at sea);

f) Transit country means any country other than the exporting or importing country across which a transfrontier movement of hazardous waste takes place or is envisaged;

g) Countries concerned mean the exporting, transit and importing countries;

h) Entity means the waste generator and any natural or legal, public or private person, acting on his own behalf or as contractor or sub-contractor (export, import, transport, collection, disposal, etc.), who owns or has the possession of the waste.

THE RESPONSIBILITIES OF THE COMPETENT AUTHORITIES IN REGARD TO TRANSFRONTIER MOVEMENTS OF HAZARDOUS WASTES

J.P. Hannequart
Consultant to Environment Directorate
OECD, Paris

SUMMARY

Consideration of current national laws indicates that need for governmental control over transfrontier movement of hazardous wastes was not foreseen by several countries. Hence, governmental control in this area varies as to extent, approach and level at which intervention may occur.

In order to adhere to international legal harmony in terms of rigor and mode of application, several actions or responsibilities will need to be assumed in future by governmental entities concerned with transfrontier movement of hazardous wastes. These responsibilities include, inter alia: collection of information; exercise of the right of refusal; issuance of licenses allowing movement of the waste; inspection methods and schedules;adoption of emergency measures in case of accident; punishment for offenses; damage compensation in certain cases.

INTRODUCTION

Transfrontier movements of hazardous wastes are a reality which must in the future be made subject to government control: this seems to be the wish of the entire international community as demonstrated in particular by:

-- the OECD Council Decision and Recommendation of 1st February 1984 on Transfrontier Movements of Hazardous Wastes;

-- the EEC Commission's proposal of 24th June 1983 for a Council Directive on the supervision and control of transfrontier shipment of hazardous wastes within the European Community;

-- the decision of the Governing Council of UNEP of 29th April 1980 on the export and disposal of hazardous chemical wastes, and the proposal prepared within this organisation on guidelines for the safe environmental management of hazardous wastes.

Effective government control over transfrontier movements of hazardous wastes presupposes that competent authorities be designated and assume certain responsibilities.

What are these responsibilities and how can they be exercised? This is the subject of the present report whose ultimate objective is to clarify the legal initiatives which remain to be taken to make government intervention effective.

The limits to the report therefore need to be underlined at the outset. It does not deal with all the responsibilities that governments might usefully take on in a spirit of far-reaching international co-operation on hazardous waste problems, e.g. steps to reduce waste production in their countries, promote R/D on clean technologies etc. The report only covers those responsibilities of the competent authorities needed for actual government control of transfrontier movements of hazardous wastes. Moreover, the typical case referred to in this report is that of movements of wastes from one country to another excluding in particular waste movements from a given country to the high seas. Firstly, Part I will review the duties of competent authorities under existing national law in regard to transfrontier movements of hazardous wastes.

Next, we shall look at the duties of competent authorities under existing international law on transfrontier movements of hazardous wastes (OECD Council Decision) and under international law now being prepared (EEC Council Directive, UNEP Directives).

Lastly, we shall refer to legal initiatives which remain to be taken if competent authorities are really to control transfrontier movements of hazardous wastes in all countries.

THE RESPONSIBILITIES OF COMPETENT AUTHORITIES UNDER NATIONAL LAW

Under national law as it stands at present the management of hazardous wastes is not dealt with specifically in all countries. Even among OECD Member countries some impose practically no government control over how waste is managed.

As concerns transfrontier movements, national law has little to say in most countries.

A summary of national provisions concerning transfrontier movements of hazardous waste (see Annex) reveals varying degrees of government intervention:

a) In regard to transfrontier movements which are subject to control

Where specific national rules concerning transfrontier movements of waste are found, it is rare for imports, exports and transit to be dealt with separately. Table 1 summarizes the situation.

In the absence of specific legal provisions it should nevertheless be noted that some transfrontier movements of hazardous waste may be subject to rules under the ordinary law applicable to movements and carriage of hazardous wastes, or of dangerous goods in general.

Table 1

Country	Specific legal provisions relating to,		
	Imports of hazardous wastes	Exports of hazardous wastes	Transit by hazardous wastes
Germany	yes	proposed	proposed
Austria	yes	no	no
Belgium	yes	yes	no
Canada	proposed	proposed	no
Denmark	yes	no	no
United-States	no	yes	no
Finland	yes	yes	no
France	yes	no	yes
Ireland	yes	yes	no
Italy	yes	yes	no
Luxembourg	yes	yes	yes
Norway	proposed	proposed	no
Netherlands	yes	yes	no
United-Kingdom	yes	yes	no
Sweden	proposed	yes	no
Switzerland	yes	proposed	proposed

Thus, the licence that the carrier of hazardous wastes must have in certain countries (Germany, Japan, Sweden, etc..) indirectly concerns transfrontier movements.

b) In regard to types of control to be exercised

Where transfrontier movements of hazardous wastes are subject to specific regulation, this sometimes takes the form of a real prior authorization system.

Such a system is found:

-- As regards imports of hazardous wastes, under German, Austrian, Danish and Finnish law;

-- As regards exports of hazardous wastes under Finnish and Swedish law;

Nevertheless, in some cases, the rules applicable involve a mere prior and/or ex post facto declaration.

19

Thus,

-- Under French law, import of and transit by hazardous waste must be declared in advance (and the authorities may object within a certain period);

-- Under Belgian law, exports must be declared ex post facto, within eight, days and imports within one month;

-- Under American law, exports must be notified four weeks before the first consignment in each year;

-- Under U.K., Irish, Luxembourg and Dutch law, importers and exporters must inform the authorities via a trip ticket system (if such systems do not give the authorities power of authorization it nevertheless seems that they can object to certain movements of wastes where safety requirements are not met).

In some countries, there is also a system of trade licences applicable to importers and/or exporters either as such (Belgium) or as "collectors" or "carriers" of hazardous waste (Germany, Italy, Japan, Luxembourg, Norway, Netherlands, Sweden, etc..).

c) In regard to information to be provided

National law on transfrontier movements of hazardous waste is based on the need for certain data to be transmitted to the authorities for information or for the issue of a licence. Information to be transmitted is not identical in all countries; but in general terms data seem to be required concerning:

-- the characteristics and quantities of hazardous wastes involved,
-- the identity of the transactors concerned,
-- the arrangements for transporting the waste,
-- the final destination of the waste.

This information is sent, depending on the country, to one or more authorities responsible for surveillance [see (d) below].

Where an authority does have power of authorization, its decision will depend on various items of information and their veracity.

For example, an import licence for hazardous wastes in Germany requires, inter-alia:

-- a certificate of capacity to handle the wastes, from a disposer,
-- a carriage licence,
-- evidence of insurance.

In Belgium, trade importers of poisonous wastes must, among other things, produce in advance evidence of financial security and of a commitment to take out standard-form insurance.

Is agreement by the authorities of the country of destination a decisive element in national export licence rules for hazardous wastes? Apart from Swedish law, the answer seems to be no. In the United States and the Netherlands, the authorities are nevertheless required to notify the authorities of the country of destination. (A reciprocal information treaty between the Benelux countries may also be mentioned).

To what extent do the competent authorities for transfrontier movements of hazardous wastes have discretionary power to refuse authorization?

Most national systems give no precise answer to this question.

Under French law, the authorities responsible for authorising imports must make a purely "objective" decision based on the facts of the case having regard solely to considerations of public health and protection of the environment.

Under other national legal systems similar limits to the discretion of the authorities are perhaps to be found having regard to the spirit of the legislation. Under Swedish law it is provided that exports can only be authorized for waste treatment abroad which is markedly better than that offered by the public company "SAKAB".

d) As regards the distribution of powers

The level or levels of government concerned vary from country to country.

In outline, the main responsibilities in regard to the control of transfrontier movements of hazardous waste are distributed as indicated in Table 2.

This table obviously gives no more than a highly simplified idea of the actual situation. In particular, it only refers to authorities whose basic responsibility is to receive notifications or issue licences. In practice, various specific controls are exercised by other authorities.

Firstly, customs authorities may be required to make various checks on transfrontier movements of hazardous wastes. French law provides for special customs formalities in connection with imports of wastes. German, British and Irish law involve the customs authorities through their trip ticket systems.

Secondly, authorities with general police powers and some specially constituted police authorities are normally responsible in all countries for carrying out health or public security checks. They may in particular be called upon to take emergency measures in the event of accident or danger arising from transfrontier movements of hazardous wastes.

Table 2

Basic government responsibilities in regard to transfrontier movements

Country	Central	Regional	Departmental	Local
Germany	no	yes (in some Länder)	yes (in some Länder)	(yes in one Land)
Austria	no	yes (Provincial government)	no	no
Belgium	yes (Ministry of Labour)	yes	no	no
Denmark	no	no	no	yes
United States	yes (EPA)	no	no	no
Finland	yes (Ministry of the Environment)	no	no	no
France	no	no	yes (Inter-departmental directorate for industry)	no
Ireland	no	no	no	yes
Italy	no	yes	yes	no
Luxembourg	yes	no	no	no
Norway	yes (national authority responsible for pollution)	no	no	no
Netherlands	yes (Ministry of the Environment-Ministry of Transport)	no	no	no
United Kingdom	no	no	yes (in Wales and Scotland)	yes (in England)
Sweden	yes (National Environmental protection agency)	no	no	no
Switzerland	yes	yes	no	no

As an example we may mention some provisions of Belgian toxic waste legislation:

-- "Without prejudice to the duties of officers of the criminal investigation department, civil servants, officials and persons responsible for supervising enforcement of the toxic waste Act..." shall include "officials and staff of the transport authorities as regards carriage by road and rail" and "officials and staff of the customs authorities.

-- "Officials and staff concerned... may in the performance of their duties... carry out any examination, checks and enquiries and obtain any information that they deem necessary to ensure that the law and regulations are complied with".

-- "Officials and staff concerned... shall be entitled,

  i) To give warnings;

  ii) To specify a period within which the offender must regularise his position;

  iii) In the event of an offence being committed, place under seal or seize, even where the holder is not the owner, the toxic waste and the means of transport used to commit the offence;

  iv) In the event of an offence being committed, to establish an official record which shall be deemed authentic subject to proof of the contrary".

-- "The governor of the province in which toxic wastes have been abandoned may have them packaged, seized or destroyed, neutralised or disposed of. He may take similar steps in regard to toxic wastes... which are being transported, imported, exported or are in transit without complying with the regulations or the relevant licence".

-- "the governor of the province and the Mayor of the commune in which are situated toxic wastes liable to constitute a serious danger may order such wastes to be transferred to such place as they or the Minister responsible for Labour shall designate... the same authorities may call upon the armed forces, the gendarmerie and the emergency services to ensure the removal and transport of toxic wastes and the safety of such operations; in such case, they shall make such application to the Minister for National Defence and to the Minister of the Interior who shall immediately take the appropriate measures".

THE RESPONSIBILITIES OF THE COMPETENT AUTHORITIES UNDER INTERNATIONAL LAW

1. OECD Council Decision and Recommendation on Transfrontier Movements of Hazardous Wastes

On 1st February 1984 the OECD Council decided "that Member countries shall control the transfrontier movements of hazardous waste and, for this purpose shall ensure that the competent authorities of the countries concerned are provided with adequate and timely information concerning such movements".

This decision creates a legal obligation bearing in the first instance on States. However, fulfilment of the obligation naturally implies responsibilities on the part of various State authorities.

What are these responsibilities?

In all the countries concerned (exporting, transit or importing countries) it appears firstly that authorities must be charged with collecting information.

In accordance with the OECD Council Decision such information is required to be "adequate". And if reference is made to the principles set out in the Recommendation, the authorities seem required at least to check on the following information:

-- The origin, nature, composition and quantities of waste intended to be exported; the conditions of carriage;
-- The nature of environmental risks involved;
-- The type of disposal and
-- The identity of all entities concerned with the transfrontier movement or the disposal of the waste.

All such information is to be "timely". It follows by implication (in addition to a pre-notification principle) that authorities should have power to object to certain movements. Among the principles whose adoption is recommended by the Council of the OECD, we find the following: "countries should adopt the measures necessary to enable their authorities to object to or, if necessary, prohibit the entrance of a consignment of hazardous waste into their territory, for either disposal or transit, if the information provided is insufficient or inaccurate or the arrangements made for transport or disposal are not in conformity with their legislation".

It will be noted that the power of objection or prohibition to be given to the competent authorities of the importing country should not be an arbitrary one but should form part of a specific legal framework.

Moreover, it is quite clear from the OECD Council decision on transfrontier movements of hazardous waste that a "flow of information" is intended to pass from country to country. Transactors within one country could well be required to send information and reply directly to the authorities of other countries concerned. But government supervision would also be possible based on duties of notification between States to be fulfilled by the authorities themselves.

In any event, it appears that direct collaboration between the competent authorities of the countries concerned is highly desirable. Thus, if an authority in one country decides, in accordance with its legislation, to object to the import or transit of waste, the authorities of the exporting country should be informed accordingly so as to make sure that the transfrontier movement in question is not initiated at all. This is the spirit underlying the international norms formulated by the OECD (as confirmed in particular by paragraph 8 of the Recommendation).

In the same spirit, it appears that the authorities responsible for transfrontier movements of hazardous waste should be instructed to ensure that carriage and disposal of such waste takes place in accordance with the laws and regulations applicable in the countries concerned. This should include in particular the requirement for all entities involved in transport of disposal to obtain due authorization. (Paragraph 2 of the Recommendation).

2. EEC Council Directive on the Supervision and Control of the Transfrontier Shipment of Hazardous Wastes

In 1982, the EEC Commission drafted a proposal for a Directive on the supervision and control of transfrontier shipment of hazardous wastes. This proposal was based on the principle of prior notification and on a consignment document. At its meeting on 8th June 1983 the European Parliament adopted a resolution inviting the Commission significantly to amend its proposal so as to make it more detailed and more constraining.

As a result, on 24th June 1983 the Commission submitted to the Council an amended proposal for an EEC regulation on the supervision and control of transfrontier shipment of hazardous wastes within the European Community. This last proposal was not adopted in its entirety by the Council on 28th June 1984 but the Council did nevertheless indicate its agreement to a proposal for a directive in this connection.

Under the definitive proposal for a directive, governments will in the future have the following duties:

1. The competent authorities of Member states concerned by the shipment or transfer of hazardous wastes should be notified in advance (on a model form) of:

   -- the origin and composition of the waste including the identity of the producer,
   -- arrangements made in relation to the route to be followed and transport insurance for damage caused to third parties,
   -- measures provided in regard to the safety of the transport,
   -- the terms of the contract with the consignee of the waste.

2. The competent authority of the Member country of destination (or last Member country through which a shipment passes) should send an acknowledgement to the possessor of the waste or object to the shipment, not later than one month after receipt of the notification, with a copy for the consignee of the waste and the competent authorities of the other Member countries concerned -- any objections being justified on the basis of legislation or regulations relating to protection of the environment, security and public order or protection of health, in accordance with the provisions of this directive, other community instruments or international conventions which the Member state concerned has concluded in this field prior to notification of this directive.

3. The Member country of dispatch and the Member country or countries of transit, will have 15 days following the notification to impose, where appropriate, conditions on the transport of waste within their national territory -- these conditions, which must be notified to the possessor of the waste, with copies for the competent authorities of Member states concerned, are not to be stricter than those imposed on similar shipments effected entirely within the Member state concerned and in compliance with existing Treaties.

4. The competent authorities of the Member states concerned and the third countries should receive copies of the transport document

(drawn up in accordance with a model specified by the directive) before the shipment is carried out, and should receive a final copy of that document within 15 days from receipt of the waste.

5. Where waste leaves the Community, for the purposes of disposal outside the Community, the customs service of the last Member country through which the shipment passes should forward a copy of the transport document to the competent authority in that Member state.

As regards the scope of these rules of community law, it should be noted that the following are excluded:

a) Chlorinated and organic solvents;

b) And, on the basis of compliance with simplified legal provisions, waste, debris, mud, cinders and non-ferous metal dust intended for re-use, for regeneration or recycling on the basis of contractual agreement concerning such operations.

3. Draft UNEP Guidelines for the environmentally sound management of hazardous Wastes

As part of the United Nations Environment Programme, draft guidelines for the environmentally sound management of hazardous waste are currently being prepared. For this purpose, an expert group has in particular formulated the following principles of particular relevance to transfrontier movements of hazardous wastes.

1. The exporting country should normally ask the importing country for its express consent, but states may, under bilateral or regional agreements, agree to a tacit approval procedure;

2. An adequate period of time (which may vary from one case to another) should elapse between the notification and the export to enable the importing state to examine the proposed transfrontier shipment;

3. Exporting states should be certain that waste can be properly disposed of in the territory of the importing state before exporting any substance;

4. States of transit and importing states should be entitled to refuse to accept a transfrontier shipment.

4. International Conventions on the International Transport of Dangerous Goods

For the record, mention should be made of a number of international agreements concerning the transport of dangerous goods, in particular:

-- ADR -- "European agreement concerning the international carriage of dangerous goods by road",

-- RID -- "International regulations concerning the carriage of dangerous goods by rail",
-- IMDG -- "International maritime dangerous goods code".

All these regulations to some extent deal with problems arising from transfrontier movements of hazardous waste: they put certain information at the disposal of the authorities enabling them to exercise a degree of control and supervision.

LEGAL INITIATIVES TO BE CONSIDERED

The carrying out of the various government controls over transfrontier movements of hazardous wastes clearly requires appropriate regulation in all countries concerned. But as has been noted above such regulation does not always exist.

What are the essential legal provisions which need to be applied by countries to ensure real and effective public control of transfrontier movements of hazardous wastes? In other words, what rules should international law impose to this end? To make transfrontier movements of hazardous wastes subject to effective government control, it is of course first necessary that the authorities be instructed to exercise the necessary surveillance.

Short of setting up some international authority with special powers in this field, all the countries concerned have therefore to be obliged at international law to designate competent authorities.

As regards the OECD Member countries, such designation seems obligatory under the Council decision of 1st February 1984.

This being so, precisely which authorities should be empowered to control transfrontier movements of hazardous wastes?

It is clear that a degree of discretion must be left to the lawmakers in each country having regard to the general political and administrative set up.

But account should also be taken of the nature of the government responsibilities to be fulfilled.

In this last connection, authorities have to be designated in each country concerned to:

1. Collect information;
2. Exercise the power of objection to certain movements;
3. Grant trade licences;
4. Carry out checks;
5. Take emergency measures in the event of accident or hazard;
6. Punish offences and provide redress for damage caused.

Consequently, a number of authorities have necessarily at the outset to be taken into account by national legislation:

-- Administrative authorities themselves;
-- Inspection services;
-- The police;
-- The customs;
-- The courts.

We shall now proceed to review those government responsibilities which seem to us essential, while stressing the legal initiatives which underlie them.

1. The collection of information

To supervise an activity it is necessary to have certain information concerning it. In all cases, certain authorities should be charged with collecting information on transfrontier movements of hazardous waste.

Insofar as the idea is to supervise transfrontier movements "from door to door", information would seem to be required by the authorities in both the exporting and the importing country and, where necessary, in the country of transit.

In practice, legal obligations to provide the authorities with information must therefore be imposed in all entities involved in a transfrontier movement of hazardous wastes.

What information do the supervisory authorities in the various countries concerned need?

Everything will of course depend on the extent of the government supervision desired. Thus, the EEC Directive requires notification of more information than does the OECD recommendation.

In any event, it seems desirable that information be required as to the nature and quantities of wastes involved, the names of the persons concerned, transport arrangements and the final destination of the wastes.

To facilitate government control, give it a degree of objectivity and make it fairly speedy, it is in practice inevitable that a standard form should be used.

Moreover, some information has to pass from one country to another. This transfer of information between States should be ensured either through obligations imposed on individuals or through direct collaboration between competent authorities in the various countries.

All the entities involved in a transfrontier movement of hazardous wastes should be required to notify, directly or indirectly, the authorities of the exporting, importing and transit countries. The use of a standard notification form should be obligatory.

2. The exercise of the right to object

The very principle of effective government supervision of transfrontier movements of hazardous wastes implies that specific authorities are able to react in good time and where necessary give directions.

Prior notification of these authorities is therefore necessary.

Prior notification of transfrontier movements of hazardous wastes to the authorities concerned should be required in all countries. It is necessary,

-- Either that the entities involved in such movements and situated within the jurisdiction of one country be required directly to transmit information and directly to reply to the authorities of other countries concerned or;

-- That the competent authorities in one country be empowered to make contact with the competent authorities in the other countries concerned.

Government supervision does not necessarily require a system of prior notification. However, the latter is in principle perfectly possible: states have the sovereign right to organize waste management in whatever way they please... subject to applicable international law.

In this last connection, the right to refuse imports of hazardous waste seems indisputable under the UNEP proposal for directives on the safe environmental management of hazardous wastes.

However, under the OECD Council Decision and Recommendation on transfrontier movements of hazardous wastes, the right of the competent authorities to object is not recognized as being unlimited. It is envisaged that authorities may object to entry of a consignment of hazardous waste into their territory for disposal or transit "if the information provided is insufficient or inaccurate or the arrangements made for transport or disposal are not in conformity with their legislation". According to the preamble to the Decision and Recommendation transfrontier movements of hazardous wastes are justified "in order to make use of appropriate disposal facilities in other countries". Only where such purpose is absent do the competent authorities seem to be entitled to prevent a transfrontier movement of hazardous wastes.

The EEC Directive on transfrontier shipment of hazardous wastes is a little more clearly restrictive as regards the discretion of the competent authorities. The proposal provides that grounds must be given for every objection by the authorities, while providing that they may require any information to establish that disposal will preserve human health and is environmentally safe. Furthermore, a recital to the EEC proposal states the principle that an efficient and coherent system of supervision and control of the transfrontier shipment of hazardous waste should neither create barriers to intra-Community trade nor affect competition.

In any event, it seems to accord with the general spirit which should prevail at international level that the right of objection by the competent authorities be confined within certain limits.

In the same spirit, it is logical that the competent authorities of one country be required to object to a transfrontier movement of hazardous wastes where they have been informed of a potential breach of laws or regulations applicable in the other countries concerned. It is also logical for them to be required not to object to the reimport of wastes which have reached another country illegally.

3. The issue of trade licences

Insofar as the aim is to ensure that transfrontier movements of hazardous wastes respect certain rules for the protection of human life and the environment, a licensing system for entities involved in transport and final disposal of the waste is practically indispensable.

The OECD Council recommends adoption of such a system. So far as disposers are concerned it already applies in nearly all those countries which have hazardous waste legislation. A transport licence referring specifically to hazardous waste is found less frequently under national law but its general introduction is nevertheless desirable.

For the purposes of effective government control, all entities involved in the transport and disposal of wastes forming part of a transfrontier movement should be required to obtain a trade licence.

4. Inspection

Effective government control of transfrontier movements of hazardous waste clearly means that the authorities must not be content merely passively to receive information. The accuracy of the information has to be checked.

It is therefore necessary that one or more specialised services be given power to carry out "on the spot" inspections of the content of a consignment, of means of transport, of practical disposal arrangements, etc.

Authorities in all countries concerned must be given powers of surveillance and control over the actual transfrontier movement of hazardous wastes.

In order that transfrontier movements of hazardous wastes may be supervised at all times and, in particular, during transport operations, the competent authorities must in practice be able to refer to a consignment document. Such documents are at present only required by the national law of a few countries. Even then the documents vary in form and content constituting an obstacle to "international" supervision. To remedy this situation, the EEC Directive on transfrontier shipments of hazardous wastes rightly provides a model form to be used in all cases.

A standard (internationally harmonized) consignment document should be required in all countries for transfrontier movements of hazardous wastes.

Border crossing points are naturally of special importance for government control of transfrontier movements of hazardous wastes. To simplify the task of the competent authorities (i.e. the customs services) and ensure that control is effective it seems especially useful to insist that transfers of hazardous wastes are only effected at specified border crossings.

In general, effective control also makes special routes desirable for transfrontier movements of hazardous wastes.

Obligatory border crossings and special routes should be prescribed in all countries for transfrontier movements of hazardous wastes.

5. The adoption of emergency measures

Irrespective of how strict government control over transfrontier movements of hazardous wastes may be, such movements will still be liable to cause accidents and create hazards. In such cases, it is normal for government to take action.

In particular, transport accidents are likely to occur. In that case, police authorities are required to take the necessary emergency measures.

In order that appropriate steps may be taken, it is clearly essential that information on hazards arising and possible remedies be available. In view of possible urgency, it is desirable that such information be available on the site of the accident. It is thus essential that all hazardous waste being transported be accompanied by documents containing instructions to be followed in the event of accident.

Regulations on transfrontier movements of hazardous wastes should in all cases include the requirement that consignments be accompanied by instructions to be followed in the event of accident or hazard.

6. Punishment of offences and compensation for damage

Government control of transfrontier movements of hazardous wastes implies that such movements be subject to certain mandatory rules, i.e. that offences be punishable by law.

In the event of failure to comply with the prescribed rules, the competent authorities (i.e. the police and the courts) are required to apply the relevant criminal penalties.

It is for national legislation to specify fines and sentences of imprisonment applicable to breaches of regulations on transfrontier movements of hazardous wastes.

Furthermore, since transfrontier movements of hazardous wastes are always liable to cause civil damage, government will inevitably make provision for compensation. The competent courts must at least be empowered to decide cases on the basis of normal civil liability rules (and thus require the party liable to make reparation for damage caused).

Over and above the normal liability rules, it is desirable for specific liability rules and special financial machinery to be applied. Reparation of damage to public health or the environment caused by hazardous waste, in particular in case of transfrontier movements, is inadequately covered by the normal fault-based civil liability rules. At any rate this is the view taken by law-makers or courts in some countries which apply the rule of strict liability based on risk and/or concentrate liability on the waste generator and/or apply compulsory insurance and guarantee fund provisions.

In the absence of international harmonization there is, to say the least, unfortunately considerable uncertainty on the civil liability issue. To ensure appropriate reparation of damage caused by transfrontier movements of hazardous wastes, special civil liability rules (harmonized internationally) should be applied in all countries concerned.

Annex

SUMMARY OF NATIONAL LAW ON TRANSFRONTIER MOVEMENTS OF HAZARDOUS WASTES

Rules of German law

All importers must fill out a standard form and obtain a declaration from the disposer (certifying that he is ready to accept waste of the volume and nature described and concerning the capacity of his installations to carry out the disposal without damage).

The characteristics and composition of the wastes in question must be proved by analysis save where the competent authority waives this requirement.

The competent authority may require additional analyses and other documents, in particular a licence to transport hazardous goods on roads and evidence of insurance.

The "competent authority" is determined by the legislation of the Land concerned. (This is the Ministry in Bavarian, Berlin, Bremen, Hamburg and Schleswig-Holstein, the environmental protection agency in the Saarland, district authorities in Baden-Württemberg Hessen, Nordrhein-Westfalen and Rheinland-Pfalz, and the municipal authorities in Niedersachsen).

Licences are granted for a single import consignment or for a fixed period where the same type of waste is to be transported in the same way on several occasions.

Eight copies of the licence application must be submitted including seven addressed to the competent authorities.

Where a licence is granted the importer receives five copies:

-- 1 for himself;
-- 1 to be handed over by the carrier to the customs services when crossing the frontier;
-- 1 to be handed by the carrier to the disposer who must send it by registered post to the licensing authority;
-- 1 for the carrier, to be sent by registered post;
-- 1 for the applicant where sent by registered post by the disposer.

The import of special wastes is subject to a charge of between 100 to 10 000 D.M.

As regards export of and transit by hazardous waste there is a proposal for a third amendment to the Waste Disposal Act whereby shipments would require express authorisation.

Rules of Austrian law

All imports of special and hazardous wastes are subject to authorization by the head of the government of the province in which disposal is planned. Authorization is not necessary for the transit of waste.

Rules of Belgian law

Persons importing for purposes of disposal and persons purchasing with view to export must be authorized by the Minister for employment and labour (and must for this purpose provide evidence of financial sureties and an insurance contract).

The licence application is made to the health and industrial medicine authorities.

Where a trade activity is involved, the application is submitted to an approval board for its opinion.

Documents proving that the necessary licences have been granted must accompany poisonous wastes in the course of transport.

Exports must be declared within eight days and imports within one month.

The charge payable for a single operation is 1,000 francs and in the case of trade activities 10,000 francs.

In the Flemish and Walloon regions there are also two Orders prohibiting any disposal of foreign wastes unless authorized by the regional authorities concerned.

Rules of Danish law

Firms wishing to import chemical waste must notify the appropriate municipal authorities at least one month in advance.

Rules of American law

Exports of wastes must be notified to the EPA four weeks before dispatch of the first consignment in each year. On the basis of this notification the EPA informs the country of destination via the Department of State.

As regards PCB's the Toxic Substances Control Act will apply and no import or export is possible without specific international memorandum of understanding.

Rules of Finish law

Exports and imports -- save in exceptional cases determined by the Ministry for the environment -- must be notified to the Ministry at least 30 days in advance.

The Ministry may refuse authorization if it has reasons to believe that the waste will not be transported or treated properly from the environmental standpoint or as regards arrangements for the management of the waste.

The Ministry may require that the person who submitted the notification forward a report within a specified period confirming that the waste has in fact been imported or exported. The Ministry will transmit notification of approved exports to the authorities of the country of destination.

Rules of French law

Imports of hazardous wastes are subject to the prior dispatch of a registered letter to the Commissioner of the Republic for the department concerned and to the appropriate Interdepartmental Directorate for Industry. This letter must be signed jointly by the producer, the carrier and the importer; it must among other things specify the proposed frontier crossing point.

The waste in question can only be imported two months after the date of the acknowledgement of receipt of the letter. Authorization may be refused having regard to the technical conditions but solely on the ground that these do not adequately protect human safety or the environment.

The appropriate authorities in the department keep informed the customs office which is to deal with the formalities relating to the import of the waste.

In the case of transit from one frontier to another, the prior declaration must be sent by the carrier to the Minister for the Environment (pollution control directorate).

Rules of Irish law

When wastes are imported they must be considered as being produced at the place of entry by the importer or the person who receives them at the place of entry on behalf of the importer.

In regard to exports,

-- the producer or holder completes part A of the five copies of the consignment note, has part B completed by the collector, gives him one copy, sends two copies to the authorities of the locality, attaches one copy to the documents sent to the person who will receive the waste at the place of entry in the country to which the wastes are sent, and retains one copy;

-- the carrier must send with the waste transported one copy of the consignment note until the waste leaves the country or is embarked on a ship or aircraft for export at which time he should retain the form.

-- the producer or holder must obtain evidence that waste has arrived at its destination or must inform the local authorities.

Rules of Italian law

In regard to imports or exports consignment documents must be drafted in the language of the country of departure or arrival.

Rules of Luxembourg law

Importers and exporters of poisonous and hazardous waste are subject, as collectors, to an approval procedure.

Any importer must obtain a certificate from the foreign government stating that the waste has been duly declared and that the state in question is unable to dispose of the waste within its own territory.

On the export of poisonous or hazardous waste, sheet number 4 of the consignment document (which accompanies every shipment) must be signed by the customs and returned to the competent service by the carrier.

Copies of documents which have to be completed abroad in connection with the disposal must be returned together with sheet no 4.

Imports of waste for re-use in Luxembourg or for transit must also be accompanied by the identity document.

Rules of Dutch law

Only authorized persons may import chemical waste save where the waste is merely in transit.

Exports are subject to the same system of notification as all other transactions. The Minister responsible is charged with notifying the authorities of the country of destination where appropriate.

The Minister responsible is the Minister for the environment in liaison with the Minister responsible for transport.

Rules of Norwegian law

Under the outline legislation, a regulation is being prepared making imports and exports of hazardous waste subject to authorization by the national pollution control authority.

Rules of UK law

Any person who imports special wastes into the United Kingdom is deemed to be the producer and the place of import is treated as the place of production.

Any person who exports special wastes is treated as the disposer for the purposes of the consignment note system.

The consignment note system, which is applied to all movements of waste, is based on a form in six copies two of which are intended to provide prior and ex post facto information for the local authorities concerned.

In England the county councils and the greater London council are the authorities responsible for waste management. In Wales and Scotland responsibility lies with the district authority.

Rules of Swedish law

Exports of hazardous wastes are subject to authorization by the national environmental protection bureau, although this rule does not apply to exports by the public company "SAKAB".

An export licence will only be granted for waste treatment abroad clearly superior to that which SAKAB can provide. Agreement of the authorities of the country of destination is required.

Rules of Swiss law

The Federal Council is responsible for regulating the import, export and transit of hazardous wastes and in particular for the labelling of packages of hazardous wastes.

All imports are subject to licence.

OECD MEMBER COUNTRIES	SPECIFIC LEGISLATION ON HAZARDOUS WASTES
1. Germany	Act of 7th June 1972 amended inter alia 21st June 1976. Order of 1974, 1977, 1978 etc.
2. Australia	-
3. Austria	Act No. 186 of 2nd March 1983
4. Belgium	Act of 22nd July 1974; Royal Decrees of 1976; Regional Decree of 1983
5. Canada	(Act of 1980 on transport of hazardous materials; regulation of 19th December 1982)
6. Denmark	Act of 24th May 1972; Decree No. 121 of 17th March 1976 and No. 323 of 3rd July 1980; Act of 18th May 1983 (Act of 1973 on the environment).
7. Spain	(Catalonian Act of 24th March 1983).
8. United States	Act 94.580 of 21st October 1976, Regulation of 1980, 1981, etc.
9. Finland	Act No. 673 of 31st August 1978 amended on 30th February 1981; Order of 1979
10. France	Act of 15th July 1975; Decree of 19th August 1977; Order of 5th July 1983.
11. Greece	-
12 Ireland	Regulation No. 33 of 1982.
13. Iceland	-
14. Italy	Act No. 915 of 10th September 1982
15. Japan	Act No. 137 of 1970, amended in 1976; Decree No. 300 of 1971
16. Luxembourg	Regulation of 18th June 1982
17. Norway	Act No. 6 of 13th March 1981.
18. New Zealand	-
19. Netherlands	Act of 11th February 1976, Decree of 1977, 1980, etc.
20. Portugal	-
21. United Kingdom	Regulation No. 1709 of 1980
22. Sweden	Ordonnance No. 346 of 22nd May 1975
23. Switzerland	(1983 Environment Act)
24. Turkey	-

INTERNATIONAL LAW MEASURES TO IMPLEMENT THE PRINCIPLES
IN THE OECD DECISION ON TRANSFRONTIER MOVEMENTS OF
HAZARDOUS WASTE

Pierre M. Dupuy
Professor at the Université de Paris 2, France

SUMMARY

The paper reviews ways of giving effect to the content and implications
of the principles in the Decision of 1st February 1984 from the public
international law standpoint, and discusses how measures to be taken by States
are to be tied into their domestic legal systems and into the framework of
their international cooperation.

INTRODUCTION

The aim of this report is to examine the principles attached to the
OECD Council Decision of 1st February 1984 and decide in the light of that
examination how best to ensure effective and efficient implementation of those
principles at international law. In particular it will be seen that
implementation of several of the provisions in the Decision depends on the
prior conclusion of _agreement_ between States directly concerned by
transfrontier movements of hazardous waste. In each case, an attempt will be
made to indicate the nature of the problems encountered and the type and
content of the required agreements.

To be effective, control of transfrontier movements of hazardous waste
requires the adoption of legal measures of two quite different types:

-- Domestic law measures. Each State must extend or adapt its own
   legislation concerning the generation, storage, carriage, and
   disposal of hazardous waste, irrespective of whether the waste
   originates in that State or comes from a foreign State.

-- International law measures. Transfrontier movements by definition
   affect two or more States. they presuppose the adoption of rules of
   conduct among those States along with regular efforts to improvement
   cooperation.

39

However, although different, these two types of measures, representing two distinct areas of State action, are nevertheless closely related. The OECD Council Decision aims to institute "cradle-to-grave" international supervision of hazardous waste. Continuity of international supervision therefore presupposes that measures taken by the competent authorities in each country be coordinated. It is also necesary that the measures be mutually recognised by the different national authorities. Mutual recognition will only be possible when both sides have confidence in the effectiveness of national legislation and administrative bodies active simultaneously or at different stages of transfrontier movements of waste.

It is also clear that, without interfering with the internal jurisdiction of individual States, international supervision of transfrontier movements of waste is bound to affect national legislation nd practical steps taken in each country. In other words, implementation of the OECD Council Decision is dependent on the States concerned agreeing on a number of provisions to be included in national law and on measures to be adopted by them at international level. International cooperation here requires a minimum of coordination, if not in all cases the harmonisation of national law.

In view of the above, this report will be divided into three parts, of which the first and second are the most important:

-- Firstly, we shall look at problems which have primarily to be resolved between States and which are thus a matter for co-operation and co-ordination among the governments of each of the countries concerned.

-- Secondly, we shall consider issues involving normative action, administrative control, and affecting the legal rules bearing on private persons active in the transfrontier shipment of hazardous waste (generators, carriers, disposers).

We are not so much concerned to examine in detail all the substantive provisions applicable to such private persons, some of which are dealt with in other reports presented to the Seminar, but rather to look first of all at those which would need to be the subject of some international agreement.

-- Thirdly, we shall refer to those matters which, at a subsequent stage of co-operation among Member countries, might be the subject of agreements or decisions aimed at improving co-ordination between the action of national authorities, or at substantial harmonization of domestic law provisions applicable to private persons, at which stage there would be some unification of the law and administration concerned with transfrontier movements of hazardous waste.

CO-OPERATION BETWEEN STATES

The Decision itself consists of no more than a brief paragraph requiring Member countries to control transfrontier movements of hazardous waste and to endure that the competent authorities of the countries concerned "are provided with adequate and timely information" concerning such movements.

Only in the second paragraph, which in formal terms is no more than a Recommendation, is there a reference to the principles which are set out in a later part of the text.

To give some or all these principles binding legal force, a not always essential but nevertheless useful attribute to guarantee the consistency and legal certainty of international co-operation to control transfrontier movements of hazardous waste between two or more countries, it would therefore be necessary to incorporate the principles in an instrument binding on all parties.

Were such an arrangement to be introduced among a very small number of OECD Member countries only, the appropriate instrument could be a traditional type international treaty (in solemn or simplified form) rather than a further Decision of the Organisation. Such a treaty would offer an opportunity to clarify the content and implications of the general principles referred to.

The conclusion of an agreement between the countries concerned (whether in the form of a Decision of the Organisation or as a traditional type treaty) seems in any event particularly necessary to enable Member countries to take all domestic measures necessary to prevent transfrontier shipments of waste which violate applicable provisions in the other countries.

A second requirement for the effective implementation of the Decision seems to be agreement between the countries concerned as to exactly what waste is to be treated as hazardous (1). Although the principles annexed to the Decision mean that States should only authorise those transfrontier movements of hazardous waste which are in accordance with the legislation of the "countries concerned" (2) by such movements, any disagreement between countries as to whether or not the waste in question is hazardous or not would be likely to cause serious disruption. In particular, it is to be feared that a more lax definition of hazardous waste by an exporting country could lead to uncontrolled movement of waste across its frontiers. The objective of any agreement between the countries concerned should on the contrary be to ensure regularity, consistency and uniformity of supervision of movements of waste from the place of export right through to final disposal.

However, it would be wrong to underestimate potential difficulties in agreeing a standard nomenclature for hazardous waste. These difficulties explain in particular why such agreement is unlikely to be reached rapidly. In the interim period the countries concerned would have to agree, and apply appropriate provisions under domestic law, on an obligation to be imposed on generators of waste involved in a transfrontier shipment, i.e. before initiating any transport operation leading to disposal in another country, generators should make sure that the waste is not defined as hazardous in any of the foreign countries concerned (transit country and/or country of disposal). If this were not the case they would presumably be required to notify the competent authority of the exporting country, to which they are of course accountable. The competent authority would then take all necessary steps to ensure that the transfrontier shipment took place in accordance with the law of the countries concerned. To this end it would co-operate with the appropriate authorities of those countries, notably by notifying them of the intentions of the generator. However, in the long-term, arrangements of this type would be no substitute for the adoption of a standard list of hazardous wastes by the countries concerned.

In general, one of the most effective measures to be included in any international agreement would make authorisation of transfrontier shipments of hazardous waste subject to production by the carrier of express approval of the shipment by the competent authorities of the foreign country or countries concerned.

Once confronted with this approval by its opposite number in the other country, the national authority of the generator would have the assurance that the transfrontier movement was not contrary to foreign legislation. Production of the certificate of approval would thus release the competent authority of the exporting country from any responsibility for the shipment once it had crossed the national frontier. Such provision is thus both the simplest and most effective way of ensuring, with the minimum commitment by each of the countries concerned, compliance with the relevant legislation of each country.

Another field in which express formal agreement of the countries concerned seems necessary to guarantee the safety of transfrontier movements of hazardous waste is that of the legal rules and practical arrangements for the international carriage of the waste. Another report has moreover pointed out the importance of adopting an international transport document to accompany the waste (3). The information to be included on the transport document will in no way affect information that has to be provided in connection with prior notification of the operation for the purposes of obtaining the necessary authorisation in the countries concerned.

It is essential for the information of the various authorities concerned, notably the customs, but also of any other person involved, that the consignment of hazardous waste be accompanied by a transport document specifying in particular:

-- the description of the consignment,
-- the origin, nature and characteristics of the waste transported,
-- the nature of the risks to the environment in the event of accidental dispersion,
-- the destination of the consignment, and perhaps the type of intended disposal.

In the above list are to be found all the essential items of information which should generally be supplied to all competent authorities concerned with the transfrontier shipment of hazardous waste. In this respect the international transport document, one of the preconditions of transborder shipments, is also one of the best ways of transmitting information concerning the waste at international level, as required moreover by the principles attached to the OECD Council Decision.

But, aside from deciding on the content of the transport document, rationalisation of practical transport arrangements seems to require some form of international agreement. Technical questions such as fixing the quality of containers (or loading or packaging methods), particularly where different types of waste travel together and have to be made compatible and/or properly isolated, are difficult to reconcile with major differences in regulation from one country to another.

Closely linked to this is the adoption of international arrangements for the inspection and identification of consignments. The international transport document referred to above is not sufficient. Consignments must also be identifiable by a label clearly understandable by all concerned. This could take the form of some iconographic, non-linguistic symbol which States had agreed upon in advance and undertaken to make known in their countries and among their authorities.

Finally, still in connection with transport arrangements, and as moreover has been advocated by other current proposals (4), agreement by the countries concerned on a restricted number of border crossing points for hazardous waste would considerably facilitate controls.

Another effective measure, requiring agreement among the countries concerned, would be the designation by each country, notified to the other country, of regional and national authorities responsible for administration and supervision of transfrontier movements of hazardous waste. This provision seems all the more essential to the extent that control presupposes a system for the mutual exchange of information which is both rapid and suitable for all circumstances, including accidents (5). In each country the competent authority must know, in any given case, exactly who it should get in touch with or who it should ask to provide information required for the purposes of the above controls. The competent officials in the different countries must also be able to make reliable contact at all times with clearly defined opposite numbers (6).

One particular area of co-operation between competent authorities which well illustrates the advantage of determining exactly who these authorities are, would seem to require some special agreement or special provision in a general agreement. This is the case where situations have to be dealt with under principles 3 (c) and 9 which relate to the reimport of waste, either because "arrangements for safe disposal cannot be completed" or because "an importing or transit country opposes in conformity with its legislation a transfrontier movement into its territory and the waste has already left the exporting country"). In both cases provision should be made for direct coordination of action by the competent authorities in both countries, the country where the waste is situated and the reimporting country, so as to avoid waste being blocked at frontiers or simply "disappearing".

The principles annexed to the OECD Council Decision include a section entitled "International Pre-Notification and Cooperation", under which the two sub-paragraphs of principle 5 deal with measures to provide the authorities of countries concerned with "adequate and timely information". Principle 6 is then designed specifically to ensure that a "request for relevant information" from importing or transit countries meets with a satisfactory response.

However, as such, none of these provisions, formulated in very general terms, says anything very specific as to the practical arrangements or even as to who is to provide the information. On this last point, only in principle 5.1 is reference made to "entities" within the jurisdiction of the countries concerned as being the persons required to provide information for the authorities of the other countries involved. This deliberately vague wording enables those States wishing to avail themselves of international cooperation to control transfrontier movements of waste to choose the channels through which they wish information to pass. The general terms in which the provision is drafted are, however, only positive to the extent that the countries concerned do in fact reach some decision. In themselves, principles 5 and 6 are too imprecise to be effective in practice.

Two main courses of action seem to be available to countries:

a) Firstly, the "entities" concerned, in most cases private ones, would be mainly or exclusively responsible for providing foreign (government) authorities with adequate information, as detailed in principle 5.2, i.e. "this information should specify the origin, nature, composition, and quantities of waste intended to be exported, the conditions of carriage, the nature of environmental risks involved, the type of disposal and the identity of all entities concerned with the transfrontier movement or the disposal of the waste". The provider of the information will, depending on the circumstances, be the generator/exporter of the hazardous waste, the importer/disposer and/or the carrier. The government authority to which the provider of information is answerable will merely ensure that appropriate regulations are in force and check that information is in fact provided, e.g. by making authorisation of the shipment conditional on production by the generator to his national authorities of the necessary permits required abroad or other document proving that the foreign authorities have been duly notified.

b) Secondly, and also in line with principle 5.1 and, even more so, with principle 6, the government authority concerned with itself assume responsibility for notifying importing and transit countries by getting in touch directly with its foreign opposite number.

There is little point in discussing the respective advantages and drawbacks of these two courses of action at length, and they can in any case to some extent be combined (7). It is probably true that the best way of making sure that information is actually transmitted is for the authorities to undertake the task themselves. But in certain cases arrangements between States (or at least between government authorities) might result in administrative complexity which would detract from the simplicity of sending the relevant information, itself the best guarantee of efficiency. It is therefore for Member countries themselves, having regard to their circumstances, and to the nature and volume of transfrontier shipments of hazardous waste, to reach agreement on the most appropriate system.

However, what is certain is that here, as in the cases referred to above, efficiency will not result from uncertainty. Irrespective of what solution is adopted by each country concerned, it will have to be brought to the knowledge of the other countries. thus, even if each country remains free to decide who is required to provide information and to whom, coordination of channels through which the information is passed will presuppose agreement between the States concerned. It is therefore not possible here to suggest a standard treaty provision or provisions. But the need for as comprehensive a provision as possible for inclusion in an international agreement on the control of transfrontier movements of hazardous waste should be stressed, if principles 5 and 6 of the OECD Recommendation are to become effective.

CONTROL OVER THE ACTIVITIES OF PRIVATE PERSONS

The principles attached to the Council Decision keep referring to compliance by the "entities" involved with the legislation of the countries concerned. Classification of these references by subject matter would seem to give the following three main categories:

a) Firstly, under principles 2(a) and 3(a) and (b) we find requirements that the transfrontier movement of waste should be in accordance with the national legislationof the countries concerned.

-- Principle 2(a) states the general rule that entities should abstain from participation in transfrontier movements "which do not comply with the laws and regulations applicable in the countries concerned".

-- Principles 3(a) and (b) make the transport and disposal of waste dependent on the generator obtaining all necessary authorisations required in the countries concerned.

b) Secondly, under principles 7, 8 and 9 domestic legislation is dealt with from the standpoint of the possible consequences of actual or potential failure to comply with it or ignorance of it by entities involved in the transfrontier movement of waste. Thus,

-- Principle 7 deals with possibilities available to the authorities of importing or transit countries should they consider that the transport (or disposal) operation does not appear to comply with their law. They are entitled in such cases either to object to or simply to prohibit the entrance of waste into their territory.

-- Exporting countries should in such cases (priniples 8 and 9) take all practicable steps to prevent the transport operation and even where necessary authorise the re-import of waste which has already crossed the frontier in spite of the prohibition.

c) Third and last, principle 4 is something of a special case. It is aimed essentially at persuading the exporting country, when applying its legislation to shipments of hazardous waste, to refrain from any discrimination between domestic movements and those intended to cross national borders.

This means in particular that the competent authorities of the exporting cuntry should monitor those stages of the transfrontier movement which take place within their territory (e.g. packaging, specification and identification of content, loading, insurance) just as stringently as they monitor domestic movements of hazardous waste.

There is thus a series of cross references to domestic law and regulations which bears out the observation made at the beginning of this report that the legislation of all countries concerned requires a minimum of coordination, or even harmonisation, to make the rules of law to which entities are subject compatible and hence capable of being complied with. There would clearly be no point in weighing down the generator, carrier and disposer with a mass of regulations, if the resulting contradictions or complexity were to prevent realisation of the desired objective, i.e.g the rationalisation of transfrontier movements of waste so as to facilitate supervision.

Where this has not yet been done, national legislation should be made compatible. Here again this will be achieved in two complementary rather than alternative ways, namely internatioal aagreement as to the main points on which national law and administrative practice should be brought into line, followed by the application of appropriate measures by each country to give effect to such agreement.

The provisions required at international level will vary depending on the content of the relevant national legislation. Here we can do no more than mention, by way of example, some of the main areas in which it might be necessary to provide at international level for the application of domestic measures by each of the countries concerned. The areas may be divided into three main categories:

a) Firstly, having regard, in particular, to the terms of principles 2(b) and 3(b) annexed to the Decision, States will have to make compatible (although not necessarily identical) their conditions for authorisation of the international carriage of hazardous waste.

It follows from the provisions referred to above that international movements of waste require a number of national authorisations:

-- by the exporting country (principle 2(b), which may also cover necessary authorisations issued by other countries),

-- by the importing country (principle 3(d)),

-- where necessary, by the transit country (principles 2(b) and 3(b).

This being so, it is clear that generators and carriers cannot be made subject to contrdictory requirements from one country to another, unless we are systematically to discourage the lawful exportation of hazardous waste (8).

b) Secondly, it seems necessary to provide in the treaty between the countries concerned that each competent authority (designated in accordance with the procedure indicated above will be provided with all necessary information on the foreign legislation with which entities under its jurisdiction (particularly generators and carriers) will have to comply in undertaking a transfrontier movement of hazardous wastes. The competent authority might also in certain cases act as intermediary between the foreign authority and the entity in the home country for the purpose of forwarding to the latter the administrative documents required by the foreign country or countries concerned or even subsequently send such documents on to the authorities of those countries (principles 3(a) and (b)).

c) Third and last, and with particular reference to principle 4, it might even be necessary so as to guarantee non-discrimination in accordance with this prinicple, to provide at international level for the introduction in each Member country of administrative or criminal penalties for failure to comply with the legislation of the other countries concerned by the transfrontier movement of hazardous waste (9).

As part of international cooperation aimed at combining the rationalisation of controls with maximum liberty of choice for each country concerned, in order to make national legislation compatible without necessarily going as far as harmonisation or substantive unification of the rules of international law, it may be thought that provisions of the types mentioned in the two preceding paragraphs would in most cases be sufficient. However, there is no concealing the fact that a number of problems would still remain and would hamper the effectiveness of international cooperation. This could result in particular from differing national rules of private international law (applicable law and jurisdiction). It is therefore of some use to consider what new international provisions will be necessary at a subsequent stage of international cooperation.

POSSIBLE SUBSEQUENT AGREEMENTS

Over and above the issues referred to above, which essentially concern how the transfrontier movement of hazardous waste is to be undertaken, outstanding and often tricky problems relate in particular to liability for international damage caused in one country by waste from another country (10).

In this respect it is seen, particularly from reading the OECD document "Hazardous Waste Legislation in OECD countries", that in spite of the recent emergence of trends favouring uniform solutions, there are still considerable differences from country to country. Countries such as Belgium provide for channelling of liability onto the generator, including that for damage caused during carriage, destruction, neutralisation or disposal, even if he has not himself taken part in these operations. But, in several other countries, there are no specific liability rules for this type of damage which continues to be covered by the ordinary law.

In the related field of insurance, several countries such as Sweden, Germany, Belgium and the United States make authorisation of transport dependent on proof by the person concerned that he is in possession of insurance covering risk arising by reason of operations connected with the transfrontier waste. In this respect if might be useful to avoid situations in which, owing to differences in national legislation, the entities concerned have to take out separate and hence costly insurance in each country to cover what is essentially the same risk.

Another tricky question is that already referred to, of differences in national rules concerning the applicable law. It might be thought that the most appropriate solution from the compensation standpoint would be to apply the law of the place where the consequences of the damage have been felt, both as regards the law applicable to the substance and for the determination of which court has jurisdiction. This solution, which ensures equal treatment for victims of damage caused by either national or foreign waste, is far from having been adopted in all Member countries, whose detailed arrangements it is at this stage unnecessary to go into.

In view of the above, consideration might be given, as an effective way of resolving these difficulties, to the negotiation of an international treaty laying down uniform substantive and procedural legal rules for all States party to it:

-- such treaty would establish rules of liability, by for example introducing a special regime of private liability for international damage (11) channelling liability onto the generator or other persons specially authorised to export hazardous waste. It would also specify the type of insurance to be provided by such person by standardising conditions. It might also specify insurance ceilings and the setting up of private guarantee funds.

-- the treaty would also indicate, not the substantive rules of the applicable law, which would be laid down in the treaty itself, but what rules would decide which court had jurisdiction and, where necessary, on what terms court decisions could be enforced.

It will be clear that the above suggestions are close to solutions of the types adopted in international treaties already in force in the fields of nuclear liability (Paris, OEEC, 1960), the transport of nuclear materials, and damage caused by the transport of oil by sea. Without being models of universal application, these international treaties have introduced quite simple rules, the effectiveness of which, subject to certain conditions such as regular reviews of insurance and compensation ceilings, seems today to be widely accepted.

The negotiation of such a treaty, although not indispensable for the application of the principles associated with the OECD Council Decision on Transfrontier Movements of Hazardous Wastes, would certainly be the most appropriate conclusion to the cooperation among the countries concerned.

NOTES AND REFERENCES

1. See in this connection the report by H. Yakowitz in this book.

2. As defined in the appendix to the Decision, i.e. the exporting, transit and importing countries.

3. See report by J. Butlin in this book. The definition of information to be included in the trip ticket does not prejudice information to be given for purposes of prior notification in order to obtain the necessity permits from countries concerned.

4. See proposal for a Community Regulation referred to above (Article 8) which goes even further and refers to "special routes".

5. It will probably not be possible for these information arrangements (dealt with in paragraphs below) to be made exclusively by the entities concerned (generator, carrier, disposer) acting on their own.

6. The making of such contacts seems even more necessary to cope with situations such as those in principle 7, where the authorities of an importing or transit country object to entry of a consignment.

7. It may be felt that a combined approach, whereby information supplied to foreign authorities by the appropriate "entity" would as necessary be duplicated, supported or corroborated by direct communications between pre-determined authorities in the two countries (cf. above para. 10), would offer the greatest degree of security.

8. It may, on the other hand, be thought that the requirements of any one importing country in regard to waste disposal could to a large extent remain a matter for that country alone and be determined in the light of circumstances in that country.

9. It would also be important for proceedings to be made just as effective irrespective of whether the legislation violated was national or foreign. This is, however, less amenable to formal agreement.

10. International damage means damage caused in a country concerned, by waste from another country.

11. i.e. liability arising as a result of the mere occurrence of damage associated with a transfrontier movement of hazardous waste without the victim having to prove any fault on the part of the generator.

HARMONIZATION OF SPECIFIC DESCRIPTORS OF SPECIAL WASTES SUBJECT TO NATIONAL CONTROLS FOR ELEVEN OECD COUNTRIES

Harvey Yakowitz*
Environment Directorate
OECD, Paris

SUMMARY

There is no uniform international approach concerning what the properties of a waste must be in order for that waste to be labelled "hazardous". Existing national lists of proscribed waste substances are extremely varied. Yet, the chief defining mechanism for quantities of hazardous waste in a given country seems to be the particular choice of substances to be placed on such lists. Accordingly, a comparative examination of definitions of hazardous wastes used in eleven OECD countries, in the European Economic Community and in six international conventions concerning dumping or carriage of hazardous materials has been performed. This examination considered concepts used in preparing lists of hazardous wastes and factors common to such lists. The results of the comparison have been used to develop the basis for a practical cross-reference system between lists of wastes considered as hazardous in OECD Member countries. Preliminary possibilities for harmonization have also been examined.

INTRODUCTION

Virtually all industrialised nations have assumed some measure of control over wastes deemed to be "potentially hazardous" by means of national legislation and/or a set of regulations overseen by a government agency. There is no particular uniformity in the legislation in defining the term "hazardous waste". However, a set of substances is typically subjected to special rules which take into account the properties, provenance and the potential effect of such (waste) substances on man and the environment.

---

* Assigned by US Department of Commerce.

In practice, a number of countries have compiled lists of wastes deemed to be hazardous. Each such list represents a practical estimate of an approach to determining what technically valid and economically equitable criteria can be applied in order to assert that a waste shall be deemed to be hazardous if some population or ecosystem is exposed to concentrations (of the waste) above some threshold assumed to be "safe". Presumably each such list reflects environmental and economic realities in the country which has compiled the list, e.g., types of industry, need to protect groundwater, population distribution, transport networks, etc. The nature and comprehensiveness of the list effectively defines the amount of wastes which will be treated as hazardous. In turn, the types and quantity of waste classified as hazardous define total incremental costs. Ultimately, the specific list of hazardous wastes is the foundation of the entire control system for such wastes.

Since the cost of dealing appropriately with a unit weight of hazardous waste is greater than for non-hazardous waste, individual generators of such wastes will naturally seek to reduce these costs. If costs are lower elsewhere, there may be a tendency for some proportion of these wastes to cross national frontiers. Thus, governmental entities, in order to maintain adequate control over hazardous wastes are faced with certain key issues associated with transfrontier shipments of such wastes:

- The transport of any hazardous substance - waste or otherwise - is regulated by OECD Member countries in order to reduce risks to man, property and the environment. Matters need to be arranged such that risks are minimised and economically useful transport activities are not impeded.

- Countries differ in definition and monitoring of hazardous waste. These discrepancies are likely to create difficulties since all transfrontier consignments must comply with local legislation/ regulation in the country of origin, transit countries and the destination country.

- Monitoring authorities, e.g., customs officials, require some means of assurance that consignments comply with local requirements. Furthermore, this means of assurance must be rapid, comprehensible and as low in cost as practical since many shipments may occur.

In order for each of these issues to be properly taken into account, some means to identify each shipment in terms of the hazardous waste list and other identifiers of exporting, transit and importing countries would seem to be necessary.

In response to public concern and the need to maintain adquate control over large amounts of potentially hazardous wastes, many industrialised nations have enacted national legislation regulating wastes deemed to be hazardous. For example, the Waste Disposal Act of the FRG as amended in 1977 states that "waste shall be disposed in such a way that the well-being of the community is not impaired". The statute then details a set of situations dealing with hazardous wastes which must be avoided by the generator; penalties for non-compliance are also set forth. The 1975 French law, the 1976 U.S. law and the 1980 UK regulations are similar in scope. However, there is no uniform international approach as to what the properties of the waste must be in order for the waste to be labelled "hazardous". Existing national lists of proscribed waste substances are extremely varied.

DIFFICULTIES IN COMPILING A LIST OF HAZARDOUS WASTES

The listing of a substance as a hazardous waste and/or the setting of threshold levels at or above which wastes are assumed to be hazardous to man and the environment implies a risk assessment on the part of those responsible for compiling the lists. There are uncertainties associated with identifying the possible hazards of each substance listed in each environmental setting in which the waste might exist. The implication is that such wastes are likely to have an unacceptably high probability of harming man and/or the environment if special care is not taken to properly treat, store or dispose of them.

The choosing of the lists of wastes deemed to be hazardous is exceedingly difficult since long term health, environmental, legal and economic issues must be taken into account and, in the best of all possible worlds, properly balanced. The nature of this problem has been stated succinctly by Berg (1).

"The best decision is reached by estimating correctly the number of people who may benefit from the correct action, or suffer from the wrong action. An overestimate can stall useful action indefinitely, on the incentive of gathering better information before taking the plunge. Policies based on such overestimates of the population at risk are currently mistaken for scientific conservativism by lawmakers and regulators. An underestimate, on the other hand, can trigger action prematurely, so that the chance of finding the right thing to do is minimised."

A. Quantities of Wastes Deemed Hazardous

The amounts of wastes which are generated annually and which may be deemed hazardous are very large. For example, hazardous waste generated yearly in the ten EEC member countries has been estimated as 15 to 20 million tonnes (2). Exact data concerning current rates of hazardous waste generation are difficult to obtain; past generation rates are even less reliable. Several OECD Member countries have reported approximate quantities; these estimates, as well as an estimate of per capita generation of hazardous wastes, are included in Table 1. The mix of industrial sectors in a given country, and world market conditions play a role in defining per capita generation of wastes. But, the chief defining mechanism for per capita quantities of hazardous wastes seems to be the particular set of laws and regulations which have been implemented in order to provide adequate control over such wastes in each country. Under existing definitions regarding hazardous waste, Pearce has asserted that the annual increase in waste generation in both the USA and the EEC is between two and four per cent (3). If so, then the annual production of hazardous waste in the year 2000 will be fifty to one hundred per cent higher than in 1983 in the USA and EEC.

52

Table 1

AMOUNTS OF WASTES DEEMED HAZARDOUS UNDER CURRENT LEGISLATION/REGULATION
AND PER CAPITA GENERATION OF SUCH WASTES IN SELECTED OECD MEMBER COUNTRIES

Country	Annual Generation of Hazardous Wastes (millions of tonnes)	Per capita Generation of Hazardous Wastes (kg/person)
Canada	$3.29^{(2)}$	135
Denmark	$0.06^{(2)}$	12
Germany	$4.5\text{-}5^{(5)}$	80
Finland	$0.087^{(2)}$	18
France	$2^{(1)}$	38
Netherlands	$0.28^{(1)}$	20
Norway	$0.12^{(1)}$	30
Sweden	$0.52^{(2)}$	63
Switzerland	$0.093^{(4)}$	15
UK	$1.5^{(5)}$	27
USA	$264^{(3)}$	1 150

Sources:

(1)   Industry and Environment (published by UNEP) Special Issue No. 4 (1983) "Industrial Hazardous Waste Management".

(2)   OECD, Private compliance costs and public administration costs in hazardous waste management, ENV/WMP/82.3 (1st Revision); for Canada, amount is on a wet weight basis.

(3)   Announcement by US Environmental Protection Agency, June 1984. Note that recent US legislation demands steps to reduce the amount of hazardous waste generated.

(4)   Milani, B., Scharer, J., Wymann, H., Les Déchets Dangereux en Suisse : réalisations, problèmes, tendances, Office Fédéral de la Protection de l'Environnement de Suisse, Berne, May, 1982, 30pp.

(5)   National Delegation to the OECD Waste Management Group (28 March, 1984).

Experience has shown that the societal costs of inappropriate disposal of hazardous wastes can be very large. Reports of costs of over $70 million for the remedial action at Lekkerkerk in the Netherlands, estimates of $250 million for Times Beach in the USA and claims of costs of $1.85 billion to restore the areas surrounding a site near Denver in the USA have been published (4, 5, 6).

B. Quantity of Transfrontier Movement of Hazardous Waste

The OECD Waste Management Policy Group (WMPG) has collected data concerning flows of hazardous wastes between Member countries and from Member countries to disposal at sea. The exact percentage of hazardous wastes which is treated, stored and/or disposed in a place other than the country of origin is difficult to ascertain, but estimates on the order of 10 per cent have been made. Of this amount, a large portion is disposed at sea either by dumping or incineration; about one-fourth is disposed in another country.

One source asserts that "some three million tonnes [of hazardous waste] are transported across borders in Europe every year" (8). Such a figure is fairly consistent with the WMPG estimate of 10 per cent. In any event, each percentage point of transfrontier waste in OECD Europe currently means that 200 000 to about 250 000 tonnes cross one or more borders on the way to final disposition. Future trends are not obvious (7,8). But note that about 200 000 tonnes in 1982 and nearly 500 000 tonnes in 1983 of such waste reportedly travelled from OECD Europe to East Germany for disposal.

So far as individual countries are concerned, data are sparse. However, note that the Netherlands exported about 35 per cent of all hazardous waste generated in that country in 1981 and 1982. Germany shipped about 25 per cent of waste generated in that country through the Netherlands for disposal at sea in the years 1980, 1981 and 1982. In 1982, the Federal Republic of Germany shipped 141,000 tonnes to East Germany; in 1983, this figure was about 275 000-300 000 tonnes.

CHARACTERISTICS OF HAZARDOUS WASTES PROSCRIBED AS OBJECTIONABLE

Lists of presumably objectionable wastes have been compiled by several OECD Member countries. Substances appearing on these lists may be defined by:

- Type, e.g. toxic, explosive, corrosive;

- Category, e.g. gas scrubber sludges, paint sludges having an organic phase, fly ashes, pesticides;

- Technology of origin, e.g. petroleum refining, electroplating;

- Generic groupings, e.g. pesticides, solvents, oily wastes, tars;

- Specific proscription, e.g. PCB's, dioxin, lead compounds;

-   Criteria leading to proscription, e.g. extraction procedure or direct analysis yielding a predetermined threshold concentration of a substance;

-   Some combination of any or all of the foregoing items.

A summary of the general classification schemes for hazardous wastes used by twelve OECD Member countries is contained in Table 2.

Table 2

CLASSIFICATION OF HAZARDOUS WASTES

Country	Type	Category	Technology of origin	Generic grouping	Special Proscription	Applied criteria for Proscription
Austria	-	+	+	+	+	-
Denmark	-	+	+	+	+	-
Germany	-	+	+	+	+	-
Finland	(a)	+	-	+	+	-
France	-	+	+	+	+	-
Italy	-	-	-	+	+	-
Japan	-	-	+	-	+	+
Netherlands	-	-	+	(b)	+	+
Norway	-	+	+	+	+	-
Sweden	-	+	(c)	+	+	-
U.K.	-	+	(d)	+	+	(e)
U.S.A.(f)	+	+	+	+	+	+

+ = used;   - = not used.

(a)   One category is designated "Corrosives"
(b)   Explosives are listed as coming under one hazard class.
(c)   Two specific technologies (surface treatment and printing/photography) form the basis for individual classes.
(d)   Tars from refining and distilling form one of 31 classes.
(e)   Criteria for the meaning of "Dangerous to Life" as applied to the list of 31 classes are given.
(f)   In addition to the Federal statute, many states have separate laws and regulations, e.g., Louisiana, California. In the U.S., all criteria can be applied depending on the location and situation.

The summary of information contained in Table 2 indicates that all of the countries for which lists were consulted apply the criterion of specific proscription to certain substances or classes of substances. Many of these specific listings are subunits of some generic grouping. Technology of origin and classification by category are invoked by most countries as well. But, no two countries apply any of the criteria or listings in the same way. A more individual countries is included in Annex 1.

In addition to a direct listing, i.e. no lower compositional limit, the basis for declaring a waste to be hazardous depends upon compositional characteristics of the waste in certain countries, e.g. Japan, the Netherlands and the USA. Thus, waste generators have the option to analyse each waste batch to ascertain whether or not the batch passes or fails the prescribed test, i.e. is hazardous or not. Alternatively, if a listed waste is present, the generator can simply treat the batch as hazardous waste without a compositional analysis.

In cases where a single threshold composition is set at a level at, or above which, any waste batch containing a listed substance must be treated as a hazardous waste, no degree of hazard is implied. The batch either passes or fails. Thus, the choice of threshold level in such a system is likely to be at least as important in determining national total amounts of hazardous waste as is the list of substances subject to the threshold test. In practice, small generators are less likely to subject their waste to analysis than are larger generators. Small generators are far less likely to utilise on-site treatment and/or disposal facilities than are their larger counterparts. Therefore, a single threshold level may well tend to reduce total national amounts of hazardous waste generated, but may make no difference to small generators who must dispose of wastes deemed hazardous.

For the situation where several threshold levels are used to decide whether substances on the national list are hazardous or not, e.g. as in the Netherlands, the outcome may be different. In such a case, the generator may well be able to make fairly precise judgments as to cost trade-offs for controlling waste batch composition and for performing the associated compositional analysis. For example, a small generator of wastes for which a very high threshold is deemed safe may be able to afford to take steps, such as bearing the cost of chemical analysis, in order to ensure that none of his waste must be treated as hazardous.

Compositional bases for declaring a listed substance as a hazardous waste may lead to complications when transfrontier movement of wastes occurs. A waste which is simply listed in one country may also be listed, but with a compositional threshold, by another country. If export from the latter through or to the former is contemplated, some means of identifying the substance and stating that, under the regulations of the country of origin, the substance is listed but that the composition is such as to declare the waste not "special" or hazardous would seem to be necessary.

EXISTING INTERNATIONAL FRAMEWORKS DEALING WITH CLASSIFICATION
OF HAZARDOUS WASTES

A. Europeean Economic Community

In addition to the lists compiled on behalf of individual countries, the EEC has issued a Council Directive dealing with toxic and dangerous wastes (9). This Directive contains a list of certain toxic or dangerous substances and materials identified as requiring priority consideration. Table III is a reproduction of this list. Presumably, the substances listed in the EEC Council Directive will eventually be incorporated into the respective lists of hazardous wastes promulgated by Member countries of the EEC. Note that Italy formally adopted the EEC listing as of 10th September, 1982. Hence, all comments concerning the EEC apply as well to Italy.

B. International Conventions Concerning Protection of the Marine Environment

A number of Member countries have ratified major marine environment protection conventions, e.g. Oslo (1972), London (1972), Helsinki (1974), Paris (1974), Barcelona (1976). Each of these conventions contains a list of substances proscribed as hazardous if deposited in the sea. Table IV provides a listing of these substances plus a cross reference to the EEC listing shown in Table 3 and the Netherlands and UK listings discussed in Annex 1. Similarities between the EEC list and the internationally agreed list of substances controlled with respect to sea dumping are apparent. These similarities occur with respect to some of the substances on the hazardous waste lists of the UK and the Netherlands respectively. Thus, the agreements already reached with regard to sea dumping may provide some basis for further international accord on transfrontier shipments of hazardous wastes.

In addition, the Helsinki Convention (1974) concerning the Baltic Sea in particular and the Paris Convention (1974) also incorporate some additional materials into their lists of proscribed substances. In particular, the Helsinki Convention places controls on the dumping of a variety of organic substances, e.g., phenols, wood preservatives, biocides, DDT, PCB's, EDTA, DTPA, phthalic acid and all polycyclic aromated hydrocarbons and their derivatives. Both the Helsinki and Paris Conventions proscribe elemental phosphorous; Helsinki Convention also lists molybdenum and tin compounds while Paris Convention lists organophosphorous and organotin compounds.

C. United Nations Recommended Classification System

The UN Committee of Experts on Transport of Dangerous Goods has been active for more than thirty years in developing recommendations dealing with all aspects of transfrontier transport of hazardous substances. According to Shaw (10):

"Experts from Canada, France, the Federal Republic of Germany, Italy, Japan, Norway, Poland, the Union of Soviet Socialist Republics, the United Kingdom and the United States of America participate regularly, as do observers from the Netherlands and Sweden. Representatives from international organisations such as International Maritime Organisation (IMO), Central Office for International Rail Transport (OCTI), International Chamber of Commerce (ICC), International Chamber of Shipping (ICS), International Road Transport Union (IRU), and industrial organisations such as European Council of Chemical Manufacturers' Federation (CEFIC), also take part in the work as well as other such groups.

Table 3

LIST OF TOXIC OR DANGEROUS SUBSTANCES AND MATERIALS
BASED ON EEC COUNCIL DIRECTIVE (ref. 9)

The following list consists of certain toxic or dangerous substances and materials selected as requiring priority consideration

1. Arsenic; arsenic compounds.
2. Mercury; mercury compounds.
3. Cadmium; cadmium compounds.
4. Thallium; thallium compounds.
5. Beryllium; beryllium compounds.
6. Chrome 6 compounds.
7. Lead; lead compounds.
8. Antimony; antimony compounds.
9. Phenols; phenol compounds.
10. Cyanides; organic and inorganic.
11. Isocyanates.
12. Organic-halogen compounds, excluding inert polymeric materials and other substances referred to in this list or covered by other Directives concerning the disposal of toxic or dangerous waste.
13. Chlorinated solvents.
14. Organic solvents.
15. Biocides and phyto-pharmaceutical substances.
16. Tarry materials from refining and tar residues from distilling.
17. Pharmaceutical compounds.
18. Peroxides, chlorates, perchlorates and azides.
19. Ethers.
20. Chemical laboratory materials, not identifiable and/or new, whose effects on the environment are not known.
21. Asbestos (dust and fibres).
22. Selenium; selenium compounds.
23. Tellurium; tellurium compounds.
24. Aromatic polycylic compounds (with carcinogenic effects).
25. Metal carbonyls.
26. Soluble copper compounds.
27. Acid and/or basic substances used in the surface treatment and finishing of metals.

N.B. :   This list was formally adopted as the national list for Italy by presidential decree of 10 September, 1982 # 915 (Gazzetta Ufficiale Della Repubblica Italiana # 343, pp.9071-9079).

The recommendations made by the Committee are not a system of regulations. They are intended to be used as a basis for national or modal rules, with a view to achieving a fundamental level of uniformity at the world level for each mode of transport. They are continuously being reviewed, revised and augmented in the light of technical developments and the arrival on the world markets of new subtances.

The first aspect addressed by the recommendations is identification. All substances considered hazardous, waste or not, are most simply assessed through their effects, and for this purpose a classification is made into: 1. Explosives; 2. Gases, compressed, liquefied, pressurized and refrigerated; 3. Inflammable liquids; 4. Inflammable solids or those liable to spontaneous combustion; 5. Oxidizing substances and organic peroxides; 6. Poisonous (toxic) and infectious substances; 7. Radioactive substances; 8. Corrosives. This classification is further broken down into specific entries and generic entries designated N.O.S. (not otherwise specified), each identified by a number".

The UN Committee of Experts also recommends that the word "Waste" precede any UN description on a shipping document when a substance is being transported for treatment leading to disposal or directly for disposal (11). Apparently, substances intended for resource recovery operations or direct recycling are meant to be exempt from the prefix word "Waste".

The UN recommendations include classification by category, generic grouping and specific substance, but technology of origin and content criteria are not included. A number of OECD member countries wish to know the technology from which waste arises, and some member countries apply content criteria. Thus a complete transition to the UN system is not directly possible.

Even though there are approximately 3000 separate entities included in the UN Recommended Classification System (UN-RCS) (12), in many instances, exact knowledge of the specific components in waste or mixtures of waste will not be available. However, the generator should be aware of the general nature of his wastes. The UN-RCS system can accommodate, to some extent, for this situation since a number of generic or N.O.S. (not otherwise specified) code slots exist. Examples include "poisonous solids" (Code 2811), "cyanide solutions" (Code 1935), "inflammable liquids, toxic" (Code 1992), "dyes" (Code 1602) plus about one hundred others. Annex 2 contains a more complete list.

The UN-RCS system has been developed by means of international cooperation and is used throughout the world. Therefore, the UN-RCS system codes may well provide useful descriptors for many transfrontier shipments of hazardous wastes. Of special interest may be the possibility for using the UN-RCS codes to help rationalise cases where one country declares a substance to be a hazardous waste and another country does not. In such instances, a UN-RCS number and the notation that the substance is a waste may provide sufficient data for officals in transit or receiving countries to classify the substance as a hazardous waste or not.

Table 4

CROSS CLASSIFICATION OF SUBSTANCES CONTROLLED WITH RESPECT TO INSERTION INTO THE SEA WITH SELECTED COUNTRY LISTS OF HAZARDOUS WASTES

Substance	Oslo Convention 1972	London Convention 1972	Helsinki Convention 1974	Paris Convention 1974	Barcelona Convention 1976	Italy, EEC (Table 2)	Netherlands	UK
1 Organohalogens	+	+	+	+	+	+	+	+
2 Organosilicons	+	+	+	+	+	-	+	-
3 Mercury and its compounds	+	+	+	+	+	-	+	+
4 Cadmium and its compounds	+	+	+	-	+	-	+	+
5 Persistent plastics	+	+	+	+	+	-	-	-
6 Crude oil	-	+	+	+	+	-	-	-
7 Hydrocarbons	?	+	+	-	+	+	+	(a)
8 Radioactive materials	?	+	+	-	+	(a)	(a)	+
9 Acids and alkalines	+	+	+	+	+	+	+	+
10 Chemical/biological warfare materials	-	-	-	-	+	+	-	+
11 Arsenic and its compounds	+	+	+	+	+	+	+	+
12 Lead and its compounds	+	+	+	+	+	+	+	+
13 Copper and its compounds	+	+	+	+	+	+	+	+
14 Zinc and its compounds	+	+	+	+	+	+	+	+
15 Beryllium and its compounds	-	+	+	+	+	+	+	+
16 Chromium and its compounds	-	+	+	+	+	+	+	+
17 Nickel and its compounds	-	+	+	-	+	-	+	+
18 Vanadium and its compounds	-	+	+	-	+	-	+	+
19 Selenium and its compounds	-	-	+	-	+	+	+	+
20 Antimony and its compounds	-	-	+	-	+	-	+	+
21 Cyanides	+	+	+	+	+	+	+	+
22 Fluorides	+	+	+	+	+	-	+	+
23 Pesticides	+	+	+	-	+	+	+	+
24 Synthetic organics	+	-	-	-	+	+	+	+
25 Carcinogenics	-	-	-	-	-	-	+	+
26 Tarry substances	+	-	-	-	-	+	-	+

(a) Dealt with under separate laws.

60

In summary, three systems by which hazardous wastes might be classified have been promulgated by internationally recognised means, e.g. the EEC Council Directive, lists in international sea dumping conventions and the UN-RCS. None of these three systems provides a complete framework into which existing lists of hazardous wastes compiled by individual countries all can be made to fit comfortably. Nevertheless, these three systems do provide a foundation upon which such a framework might be constructed. In order to develop an appropriate framework, some direct, fairly simple means of rapidly and accurately cross-referencing existing lists of hazardous wastes compiled by individual countries with one another and with the three international lists is needed.

CROSS-REFERENCE SYSTEM FOR HAZARDOUS WASTE LISTINGS

At the request of the Waste Management Policy Group, the OECD Environment Directorate has sponsored an effort to establish the basis of a simple, practical cross-reference system for wastes listed as hazardous by one or more countries. Specifically, lists provided by the following countries were included in the first phase of the work: Austria, Denmark, Federal Republic of Germany, Finland, France, Japan, Netherlands, Norway, Sweden and United Kingdom (see Annex 1). The lists developed by the EEC (and adopted by Italy) (Table 3) and on behalf of the sea dumping conventions (Table 4) were also included. Whenever possible, a UN-RCS code number was associated with the entries on national and international lists of hazardous wastes.

As indicated in Table 2, the current lists of wastes deemed to be hazardous can be divided into several subunits. The goal is to choose the simplest cross-reference system which will accommodate every substance on every list. After considering the nature and comprehensiveness of the various lists, the following cross-referencing system was selected:

1. Generic groupings of waste were selected from the lists. Examples include "solvents and wastes containing solvents", "medical and hospital wastes". Thirty-four such items were taken from the lists; (see Table 5).

2. Subdivisions of the generic groupings were required. These items were more specific waste units, e.g. "cyanide containing liquids, baths and sludges", "paint sludges with an aqueous phase", "metallic chips and particles". These specific Categories were taken directly from the publication "Nomenclature des déchets" (NDF) issued by the Ministry of the Environment of France in September 1983. Some 100 separate categories are included.

3. In order to take into account the technology of origin of the waste, the NDF listing of Arisings encompassing over one hundred possible sources of hazardous waste was selected. Examples include "Fabrication of asbestos/cement products", "Manufacture of glue", and "Manufacture of basic plastic materials".

61

4. A list of fifty-six <u>specific inorganic substances</u> was derived from the combined lists. Examples include "mercury and its compounds", "inorganic cyanides".

5. A list of fifty-four <u>specific organic substances</u> was derived from the combined lists. Examples include "chloroform", "acetone", "benzene", "pyridine".

6. There are literally millions of organic substances which might be deemed hazardous upon discard. For practical purposes, many countries have chosen to designate classes of organic wastes as being hazardous or to require treatment as special wastes; Tables 3 and 4 as well as Annex 1 provide specific examples. In order to provide a basis for deciding whether, and in which countries, a given shipment of waste might be declared hazardous, the set of questions shown in Exhibit 1 was developed. Generators and/or shippers of waste materials containing organic substances should be able to provide replies to all questions in the set within a few minutes. Once these replies are available, reference to the tabular information shown in Exhibit 1 will enable a decision to be made concerning whether and where the waste would be deemed hazardous.

Note that the order of questions in Exhibit 1 was selected so that the simplest criteria for decision appear early; more complex questions, e.g. "Does the waste contain aromatic polycyclic compounds?", occur towards the end of the list. The order of questions was also designed to allow a determination of whether or not the waste shipment is hazardous or special in a prospective transit or destination country by having to pose as few questions from the list as possible.

All wastes on all lists available could be cross-indexed with the foregoing criteria. Certain redundancies occur, i.e. one waste from the list of a given country sometimes appears in more than one <u>Generic Grouping</u>, <u>Category</u>, <u>Arising</u>, etc. In selecting possible criteria, when reasonable doubt existed, the waste was included rather than excluded; thus, the system is slightly biased towards "false positives" rather than toward "false negatives".

Cases of a waste being listed as hazardous in some countries but not in others are easily identifiable with the cross-referencing system which has been developed. Consider a listing entitled "Arsenic and its compounds" included with the Inorganic Materials series. Neither Norway nor Sweden has chosen to mention arsenic as such on their respective lists of hazardous wastes. Nevertheless, the fact that a shipment containing arsenic and/or its compounds had arrived at a border in a shipment of waste, or a shipment declared to have no intrinsic value, might lead officials to consult the cross-reference system to determine if the shipment needed to be treated as hazardous waste. Such consultation would reveal that Japan, the Netherlands, UK and EEC treat any such shipment as hazardous waste; in most cases so does France. Austria, Denmark, and Germany make reference to specific arsenical substances.

Table 5

GENERIC GROUPINGS FOR HAZARDOUS WASTE CROSS-REFERENCING SYSTEM

1. Wastes from surface preparation and finishing
2. Solvents and wastes containing solvents
3. Oily liquid wastes
4. Paint, varnish and ink wastes
5. Sludges from metalworking
6. Solid inorganic wastes from mechanical and thermal treatment
7. Roasting, melting and incineration wastes
8. Wastes from organic synthesis
9. Liquid sludges and wastes from chemical treatments
10. Solid inorganic wastes from chemical treatments
11.   Wastes from pollution control and water purification
12. Materials and contaminated materials
13. Refuse from normal use, losses and rejects
14. Common wastes
15. Urban wastes,
16. Wastes containing biocides
17. Concentrated acid or alkaline wastes
18. Laboratory wastes
19. Tarry materials from refining and tar residues from distilling
20. Pharmaceutical and veterinarian compounds
21. Animal and vegetables fats
22. Soaps and detergent waste
23. Plastic wastes
24. Rubber wastes
25. Textile wastes
26. Material and hospital wastes
27. Drilling, cutting, grinding and rolling fluids
28. Explosives
29. Photochemical wastes,
30. Animal and slaughterhouse wastes
31. Petrochemical wastes
32. Carcinogenic substances
33. Chemical and biological warfare materials
34. Radioactive wastes.

The cross-referencing system does not deal with degree of hazard. Rather, the assumption is made that the waste being shipped out of a given country would be deemed a hazardous waste within that country or that the transport documents would indicate that the waste was listed but that this batch passed the required threshold test and so -- in the parent country -- would not be a hazardous waste.

Preliminary tests of the cross-referencing system indicate that fairly rapid translations between lists can be made with a high probability of accuracy. Further testing is necessary, but initial results suggest that a practical cross-referencing system for harmonizing various hazardous waste lists is entirely feasible.

Exhibit 1

QUESTIONS MEANT TO AID IN DECIDING IF A WASTE ORGANIC SUBSTANCE IS DEEMED HAZARDOUS
OR SPECIAL BY ONE OR MORE COUNTRIES, EEC OR SEA DUMPING CONVENTIONS

(All questions are to be applied to the waste batch. If a question can be answered "yes", then reference to the table will indicate whether (+) or not (-) the waste is deemed hazardous or special by one or more countries, EEC or Sea Dumping Conventions

Question Number	Does the Waste Batch contain organic substances whose constituents include	Austria	Denmark	Germany	Finland	France	Japan	Netherlands	Norway	Sweden	UK	EEC Italy	Oslo Sea Dumping Convention	London Sea Dumping Convention	Barcelona Sea Dumping Convention
1.	Solvent(s)	(+-)[1]	+	+	+	+	+	-	+	+	-	+	-	+	+
2.	Chlorine	-	+	(+-)[1]	+	+	+	+	+	+	+	+	+	+	+
3.	Halogens other than chlorine	(+-)[1]	+	(+-)[1]	+	+	+	+	+	-	-	+	+	+	+
4.	Cyanate(s)	(+-)[1]	-	(+-)[1]	+	+	+	+	+	-	-	+	-	+	(+)[2]
5.	Isocyanate(s)				+	+	+	+						-	
6.	Phosphate(s)				-	-	+	+						-	(+)[2]
7.	Phosphorus other than phosphate-	+	+	+	+	+	+	+	+			+	+	+	(+)[2]
8.	Biocides/Pesticides	+	-	+	+	+	+	+	+			+	+	+	+
9.	Peroxide(s)				-	-	+	+						-	(+)[2]
10.	Sulphur	(+-)[1]			-	-	+	+						-	(+)[2]
11.	Phenols	+	-	+	+	+	+	+	+			+	+	+	(+)[2]
12.	Nitrogen	+			-	-	+	+				+		-	(+)[2]
13.	Ether(s)				-	-	+	+						-	(+)[2]
14.	Acid(s)	+			+	+	+	+	+			+	+	+	+
15.	Organometallic(s)	+			+	+	+	+	+			+	+	+	+
16.	Mercury	+			+	+	+	+	+	+	+	+	+	+	+
17.	Arsenic				+	+	+	+	+		+	+	+	+	+
18.	Cadmium				+	+	+	+	+			(+-)[1]	+	+	-
19.	Copper	+		+	+	+	+	+	+	-	+		+	+	(+)[2]
20.	Amine(s)						-	-							
21.	Heterocyclic compounds containing oxygen, nitrogen, or sulphur compounds	-	-	-	-	-	-	-	+	+		-	-	-	(+)[2]
22.	Aromatic polycyclic compounds	-	-	-	-	-	-	-	+	+		-	-	-	(+)[2]
23.	Hydrocarbons and their oxygen, nitrogen or sulphur compounds	-	-	-	-	-	+	+	-	-	+	-	-	-	+
24.	Aromatic hydrocarbons	+	-	-	-	-	-	-	+	+		+	-	+	+
25.	Halogen free polymer softeners	+	-	-	-	-	-	-	-	-		-	-	-	+
26.	Any cyclic compound not taken into account by a previous question	-	-	-	-	-	-	-	-	-		-	-	-	-
27.	Any aliphatic compound not taken into account by a previous question	-	-	-	-	-	(+-)[1](+-)[1]	-	+	-	(+-)[1]	-	-	-	(+)[2]
28.	Metal carbonyl(s)	-	-	-	-	-	(+-)[1]	-	+	-	+	+	-	-	+
29.	Silicon	-	-	-	-	-	-	-	-	-			-	-	+

Notes: (1) Waste is sometimes classed as hazardous -- direct reference to national list may be necessary.
(2) If the waste is synthetic in origin, it is proscribed.

The lists developed by the EEC (Table 3) and for the various sea dumping conventions (Table 4) fit directly into the cross-referencing system. In addition, UN-RCS numbers could be assigned in about 70 per cent of the cases. Thus, existing national and international approaches to hazardous waste listings can in many cases be interchanged by means of the cross-referencing system.

A SUGGESTION FOR A GENERAL CONTROL DOCUMENT

Control mechanisms for transport, export and/or import of hazardous wastes vary in seven OECD Member countries (13, 14). However, all OECD Member countries can exercise some degree of control over imports in general. Examination of required documentation relating to imports indicates that two OECD Member countries, Spain and Turkey, require a "Certificate of Origin" to confirm the origin of the imported materials. The remaining twenty-two OECD Member countries have the legal right to require such a certificate. When required, the certificate is signed by the shipper, notarised and then confirmed by a recognised Chamber of Commerce (15).

Although special formats for this certificate are required by some countries, e.g. Japan, Portugal, all formats require the exporter to provide:

- Exporter's business address
- Carrier name and type
- Date of shipment
- Consignee's business address
- Number of items, e.g. crates, boxes, drums
- Gross and net weights in kilograms
- Description of articles in shipment
- Signature of representative of the exporter attesting to the correctness of the above information.

Countries which require a manifest for transport of hazardous waste, require similar information. Thus, by the legally available expedient of requiring a Certificate of Origin for all shipments that are to enter a country without payment of duty on the grounds of being "hazardous waste of no commercial value", a measure of control can be attained regarding such shipments.

Using this means of tracking the waste would:

- Not conflict with proposals for a uniform EEC manifest;

- Not require direct intervention by governmental authorities except Customs Officials;

- Provide a means to have records of shipments which would be an appendage to an already established record-keeping system dealing with all "Certificates of Origin";

- Not require any new legislation;

- Provide authorities access to information on transfrontier shipment of hazardous waste on whatever sampling basis they so choose;

- Maintain liability with the generator and/or the shipper and waste disposal firm;

- Provide some measures of community control since an important recognised Chamber of Commerce must be informed;

- Provide whatever information might be needed to decide whether or not to admit a shipment of waste;

- Allow shippers to utilise forms with which they are generally familiar;

- Form the basis for a uniform international manifest system without necessarily requiring export and/or import licences. (NB. Countries wishing to utilise such licences might, in fact, use the information on the certificate as the basis for issuing the licence).

- Require international agreement concerning:

  i) marks and numbers associated with the shipment;
  ii) information to be provided in the description.

With regard to marks, the UN-RCS system provides detailed recommendations regarding packaging and marking for each item assigned a code number. These recommendations might provide a reasonable foundation for agreement concerning marks.

POSSIBILITIES FOR A "UNIVERSAL LIST" OF HAZARDOUS WASTES

A description of the waste load conforming to the regulations of the country where the waste had been generated would be a minimum expectation. The inclusion, where possible, of a UN-RCS code would be very useful in indicating the nature of the waste batch to the authorities of transit and/or destination countries. In addition, the constituents of the waste would be useful as an aid to these authorities in using a cross-referencing system in order to determine whether or not the waste batch is hazardous under local regulation. If the waste load were destined for sea dumping, the cross-reference system should be capable of identifying substances prohibited from such dumping. Presumably, the cross-referencing system itself would have to be internationally agreed upon whether in folio or computer format.

The possibility of a "universal" list or more properly an internationally agreed set of tests and list of substances deemed to be hazardous wastes may be considered. Some basic challenges to development of such a list include, but are not limited to, the following:

1. A listed substance would need to be included or excluded on the basis of criteria not requiring qualitative or quantitative analytical chemistry. If not, small generators will have difficulty in complying with regulations;

2. Changes in the list would require international notification and agreement;

3. Mode of treatment, storage and/or disposal would be independent of the list. If not, countries might be faced with the necessity of revising laws/regulations governing hazardous wastes;

4. The list would probably not be all-inclusive in individual countries;

5. A benefit/cost analysis of adopting any specific list would be very difficult to perform. Thus, the equitability of any specific list would always be open to some question;

6. A mechanism to deal with wastes destined for, or exported by, countries not adopting the list would be needed.

In the absence of a universal list, a comprehensive cross-reference system embodying all existing lists is necessary. Indeed, by the criterion of encompassing all listed wastes, the cross-reference system becomes de facto the operational "universal" list of wastes deemed to be hazardous. Hence, an internationally agreed cross-reference system would be very useful in helping to maintain adequate control over transfrontier movements of hazardous wastes.

CONCLUSIONS

1. Existing monitoring systems for hazardous waste in several OECD Member countries differ in required reporting and/or manifest schemes for each country; but there is no internationally accepted definition of which properties and/or provenance of a substance designates that specific substance to be a "hazardous waste". These discrepancies are likely to create difficulties since all transfrontier consignments must comply with local legislation/regulation in the country of origin, transit countries and the destination country; Furthermore, monitoring authorities, e.g. customs officials, require some means of assurance that consignments comply with local requirements. Furthermore, this means of assurance must be rapid, comprehensible and as low in cost as practical since many shipments may occur;

2. Some international lists of proscribed wastes exist including that provided by EEC Council Directive and wastes proscribed from sea dumping by means of three international conventions. The United Nations recommended classification system which deals with transport of all dangerous substances can be applied to some extent to existing lists of hazardous wastes, the total quantity, and hence the total cost of control, of hazardous wastes in a given country is strongly dependent upon the specific list and tests required to ascertain whether a given waste is special and/or hazardous;

3. Existing lists of hazardous wastes for several OECD Member countries can be (and have been) incorporated into a fairly simple comprehensive cross-referencing system. The international lists can be (and have been) included in the system; (N.B. the cross-referencing system is expected to be published by OECD in 1985); the cross-referencing system can be used in conjunction with a simple transport document to provide adequate control over transfrontier movements of hazardous wastes. Such implementation would require little or no additional legal, regulatory or administrative activity. Costs and complexities associated with hazardous waste control would increase only slightly. The cross-referencing system can be used to help harmonize the pre-notification scheme called for by the OECD Council Action of 1 February, 1984 concerning transfrontier movement of hazardous wastes.

NOTES AND REFERENCES

1. Berg, G. G. in "Measurement of Risks", edited by G. G. Berg and H. H. Maillie, Plenum Press, New York and London, (Environmental Science Research, Vol. 21) 1981, 550 pp.

2. Klein, L., "European Community Action on the Management of Waste Products" in Proceedings of the International Conference Refiuiti Solidi, Fanghi e Materiali Residui: Rilevamento, Tecnologia e Gestione, pp 17-20, Rome 1981.

3. Pearce, P., "Landfilling: A Long-Term Option for Hazardous Waste Disposal?", Industry and Environment (Special Issue) UNEP 1983, No. 4, pp 57-62.

4. Erasmus, H., "Industrial Hazardous Waste Management in the Netherlands", Industry and Environment (Special Issue) UNEP 1983, No. 4, pp 25-27.

5. "OTA Reviews Cleanup of Missouri Dioxin", Chemical and Engineering News (USA), 19th December 1983, p. 11.

6. "Toxic Cleanup", Washington Post, 5th October 1983, p. A14.

7. Butlin, J. and Lieben P., "Economic and Policy Aspects of Hazardous Waste Management", Industry and Environment (Special Issue) UNEP 1983, No. 4, pp 11-14.

8. "Toxic Wastes in the European Community", World Environment Report, 30th March 1983, p. 3.

9. Council of the European Communities, "Council Directive of 20th March 1978 on Toxic and Dangerous Waste", in Official Journal of the European Communities, 31st March 1978, L 84/43-48.

10. Shaw, P. M., "International Legislation and the Transport of Hazardous Wastes", Industry and Environment (Special Issue) UNEP 1983, No. 4, pp 63-65.

11. Roberts, A. I., "Transportation Regulations of Hazardous Waste; USA and International Developments", in Hazardous Waste Disposal, (J.P. Lehman, ed.). Proceedings of the NATO CCMS Symposium on Hazardous Waste Disposal held in Washington D.C., October 1981, New York (1982) pp 69-76.

12. United Nations Economic and Social Council, "Transport of Dangerous Goods" (Recommendations by the UN Committee of Experts on the Transport of Dangerous Goods), Second Revised Edition, 462 pages.

13. van Veen, F., "Monitoring Systems for Hazardous Waste in OECD Countries", see report in this book.

14. European Commission, "Proposal for a Council Directive on the Supervision and Control of Transfrontier Shipment of Hazardous Wastes within the European Community", submitted to the Council 17th January 1983.

15. Unz and Company, "A How-to Guide for Exporters and Importers", 1983, 60 pages. (Available from Unz and Co., 190 Baldwin Ave., Jersey City, N.J. 07306, USA).

Annex 1

LISTS OF WASTES SUBJECT TO NATIONAL CONTROLS

AUSTRIA

In September, 1983, ONORM S2101, a provisional list entitled "Hazardous waste requiring supervision" was issued by the Austrian authorities. The list comprises twelve main waste streams and thirty-six specific waste streams; in all, there are 161 entries. The twelve main waste streams are:

. Animal and slaughterhouse wastes (Code 13);
. Mineralogical (inorganic) wastes, except metal wastes (Code 31);
. Metal wastes (Code 35);
. Oxides, hydroxides and wastes (Code 51);
. Acids, bases, industrial solutions (Code 52);
. Wastes from agriculture and pesticides including those of pharmaceutical origin (Code 53);
. Wastes from mineralogical production (oil based wastes) (Code 54);
. Wastes from organic solvents, dyes, lacquers, adhesives, cements and resins (Code 55);
. Plastic and rubber wastes (Code 57);
. Textile wastes (natural and synthetic) (Code 58);
. Other chemical wastes and synthesis products (Code 59);
. Specific hospital wastes (Code 97).

A few specific entries are as follows:

. Paint Media (Code 55508)
. Barium Salts (Code 51525)
. Ethyl Acetate (Code 55302)
. Mercury Batteries (Code 35324)
. Salt Bath Wastes (Code 50511)

DENMARK

A listing published by the National Agency of Environmental Protection entitled "Types of chemical waste covered by Danish regulations on the disposal of chemical waste", dated 8 /IX /80 was available. Five major types of waste are included, viz.:

1. Animal and vegetable fats.

2. Organic compounds containing halogens.
3. Organic halogen-free compounds.
4. Inorganic compounds.
5. Other wastes.

For each of these types, examples were given such as "solid residues from organic synthesis, containing organically bound halogen and/or sulphur" (Type 2), "tar and anti rust oils" (Type 3), "waste from the production and distribution of chemical pesticides" (Type 5).

GERMANY

A "waste catalogue" (Informationsschrift Abfallarten) of about 570 types of waste was promulgated by the German Ministry of Interior in 1975. In 1977, a group of 38 of these wastes was identified as hazardous or "special" waste. The classification system consists of a five digit code of which the first is the "Major Group", the second is the "Group" and the third is the "Sub-group". The remaining two digits specify the specific substance and its probable technology of origin. For example, Code 3 represents general mineralogical wastes, Code 35 is metal wastes, 355 is metal sludges and 35503 is lead containing metal sludge which may arise from lead smelting operations or electrolytic operations. Germany intends in 1984 to expand this list in cooperation with the German States (Länder). The new list will contain approximately 330 wastes deemed special or hazardous. The Austrian system is modelled upon that of Germany; over 80 per cent of the wastes listed by Austria are identical to those of Germany.

FINLAND

The Waste Disposal Act (1978), the Planning and Building Act (1958), the Public Health Act (1965), the Water Act (1961), the Air Pollution Control Act (1982), and the Act on the Prevention of the Pollution of the Sea (1979), form the basis for the organisation and supervision of waste disposal in Finland. All of these Acts incorporate planning and notification procedures, and the obtaining of permits, related to tips, waste treatment sites, or waste treatment in general. Because of the general dispersion of environmental administration in Finland the supervision according to these Acts fall under the responsibility of several ministries.

The central regulations for waste disposal are set forth in the Waste Disposal Act. By looking at the regulations of this Act, a clear picture of how waste disposal is organised, executed, and administered will emerge. The Waste Disposal Act applies to all household and industrial wastes that are not discharged into water courses or sewers or into the air.

The general objectives of the Waste Disposal Act state that waste disposal is to be carried out in as far as possible so that waste is either recycled or otherwise made beneficial use of, and so that waste does not cause any damage to the environment. Waste disposal itself is taken to mean the collection, transportation, reception and storage of waste, the rendering harmless of waste, and other comparable treatments.

Waste means all objects and substances disposed after use, and of little or no value. Also considered waste are other objects or substances which have been collected or been brought to places reserved for waste, for transportation, storage, rendering harmless or other treatments.

Problem wastes are defined separately in the Act (Paragraph 4). Problem wastes are those wastes which, because of their poisonous or other qualities are difficult to render harmless or otherwise treat, and those wastes which are otherwise very harmful to the environment. According to a ruling of the Ministry of the Interior, No. 576/79, Paragraph 1, problem wastes are:

1. Wastes containing oil.
2. Wastes containing solvents.
3. Wastes containing corrosives, such as acid or alkalies.
4. Wastes containing antimony, arsenic, mercury, silver, cadmium, cobalt, chromium, copper, lead, manganese, nickel, zinc, thallium, or tin.
5. Wastes containing inorganic or organic cyanides or isocyanates.
6. Wastes containing organic halogenated compounds, such as polychlorinated biphenyls (PCB).
7. Wastes containing phenols.
8. Wastes containing biocides or (wood) preservatives.
9. Wastes containing medications or drugs, or the raw materials used in their manufacture.
10. Other wastes, which because of type and property are comparable to the above.

According to the same ruling, however, wastes on the above list that appear in small quantities or in small concentrations, or wastes which will not spread throughout the environment in a harmful manner are not considered problem wastes if they may, without danger and without causing harm or damage, be collected, transported, and treated along with other wastes.

According to the Act the administrator of real estate is considered the waste producer. The administrator is either the owner or the lessee of real estate. Regulations in the Waste Disposal Act which apply to administrators of real estate apply equally to those who maintain roads, administrators of railroads, those who maintain small boat and ship harbours, and those who maintain outdoor exercise trails. The responsibility for disposal of problem waste is partially laid further directly onto the producer of the problem waste (a producer of problem waste may not be the same person as the administrator of the property).

FRANCE

The French system lists general types of waste, e.g. waste motor oils, cyanide bearing liquids and sludges as well as a variety of technologies of origin for such types of wastes. Some twenty-seven types of waste are

72

classified as "special" under terms of Article 8 of the Law of 15 July 1975 (75,633) governing waste disposal and materials recovery. The decree of 19 August, 1977 (77,974) sets forth five major classes of hazardous wastes:

1. Wastes containing 27 proscribed substances, e.g. PCB, solvents (Table A1.1 contains a full list);
2. Radioactive waste;
3. Paint, waste oil and hydrocarbon sludges;
4. Wastes from certain technologies, e.g. coke production;
5. Wastes from metal finishing activities.

Specific wastes are identified by type (category) of which 100 are listed and technology of origin of which 141 are listed; a two part Code employing six digits is used. The codes are read from a master list published by the Ministry of the Environment.

Table A1.1

FRANCE: LIST OF SPECIAL WASTES
(Decree 77-974 of 19th August 1977)

Wastes containing:

Asbestos	Nickel and its compounds
Antimony	Phenols and Phenol Derivatives
Arsenic and its compounds	Lead and its compounds
Barium and its compounds	PCB
Berylium and its compounds	Selenium and its compounds
Cadmium and its compounds	Aromatic solvents
Hexavalent chromium	Chlorinated solvents
Trivalent chromium	Inorganic sulfur bearing compounds
Copper and its compounds	Organic sulfur bearing compounds
Cyanides	Thallium and its compounds
Tin and its compounds	Titanium and its compounds
Fluorides	Vanadium and its compounds
Isocyanates	Zinc and its compounds
Mercury and its compounds	Radioactive substances
Molybdenum and its compounds	

Wastes composed primarily of paint sludges, hydrocarbons, manure

Wastes originating from:

Petroleum Refining Activities
Coking operations
The Chemical Industry
Pharmaceuticals and Phyto-Pharmaceuticals
Laboratories
Activities of Surface Treatment Workshops

JAPAN

The Waste Disposal and Public Cleansing Law divides wastes into two categories: domestic wastes and industrial ones. There are 19* materials included in the latter. Hazardous wastes are defined under the same Law. They are cinder, slag, dust, waste acid and waste alkali with some harmful substances exceeding the relevant criteria, which differ depending on the forms of final disposal and the kinds of wastes. Table A.1.2. shows the specified substances and their respective criteria. Wastes to be landfilled are subjected to elution tests using pH adjusted water as solvent; an elution period of six hours is used. As for wastes to be dumped into the ocean, there are two methods of testing: elution tests for water-insoluble wastes and content tests for water-soluble wastes. In addition, ocean dumping of the following wastes is prohibited.

1. Phenol-bearing sludge, waste acid and waste alkali

2. Waste oil

Liquid wastes such as waste acid, waste alkali and waste oil may not be landfilled.

Certain technologies of origin are associated with the designated toxic substances appearing in Table A.1.2. Seventy major technologies of origin such as those of inorganic pigment manufacturing industry, methane derivative manufacturing industry, pharmaceutical manufacturing industry, etc. are involved. For example, the synthetic dye manufacturing industry is listed as having the potential to produce organic chlorine bearing wastes, cyanide bearing wastes, hexavalent chromium bearing wastes, copper bearing wastes and zinc bearing wastes. These major technologies are futher sub-divided into various unit operations, e.g., synthetic dye manufacturing has, as a sub-unit, centrifuge operations. For the purpose of this Annex, the total number of listed technologies of origin associated with the specified substance will be noted, i.e., for arsenic and arsenic compounds, some eighteen major technologies of origin are listed.

NETHERLANDS

The Chemical Waste Act of 1977 which was implemented in 1979 is the main legislative basis for hazardous waste management. Proscribed substances are listed in the schedules to the Royal Decree of 26 May 1977 governing implementation of certain sections of the Chemical Waste Act. The list was established on the basis of varied criteria including potential for harm to man and his environment, persistence, cumulative effects and toxicity.

---

* Included here are cinder, sludge, waste oil, waste acid, waste alkaki, waste plastics, slag, dust, etc.

CONTROLLED WASTES IN JAPAN

Disposal method	Landfil		Ocean Dumping	
Wastes / Specified Substances	Slag(1) Sludge Cinder(1) Dust(1)	Slag(1) Water-insoluble inorganic sludge Cinder(1) Dust(1)	Water-soluble Inorganic sludge Organic sludge	Waste acid Waste alkali
Testing methods	Elution tests	Elution tests	Content tests	Content tests
Alkylmercury Compounds	ND(2)	ND(2)	ND(2)	ND(2)
Mercury and its Compounds	not more than 0.005mg/1	not more than 0.005mg/1	not more than 2mg/kg	not more than 0.05mg/1
Cadmium and its Compounds	0.3	0.1	5	1
Lead and its Compounds	3	3	50	10
Organic Phosphorous Compounds	1	1	5	1
Chromium (VI) Compounds	1.5	0.5	25	5
Arsenic and its Compounds	1.5	0.5	25	5
Cyanides	1	1	5	1
PCB	0.005	0.003	0.15	0.03
Organic chlorine Compounds	-	40mg/kg sample (3)	40	2
Copper and its Compounds	-	3mg/1	70	15
Zinc and its Compounds	-	5mg/1	450	90
Fluorides	-	15	1 000	200

(1) Excluding Alkylmercury compounds, Organic phospherous compounds, Cyanides and Organic chlorine compounds.
(2) not detectable.
(3) content tests.

The schedules are sub-divided into classes of substances designated A, B1, B2, C, D respectively; there is also a list of nine technologies of origin (processes), the wastes from which are presumed to fall under the requirements of the Chemical Waste Act. Finally, there is a list of special exemptions.

Waste is not deemed chemical waste if the content (concentration) of the listed substances is less than:

12 substances comprise Class A: 50mg/kg
(Examples: As, Hg, carbonyls)
33 substances comprise Class B: 5 000mg/kg
(Examples: asbestos, organic halogens)
16 substances comprise Class C: 20 000mg/kg
(Examples: chlorates, Ba)
20 substances comprise Class D: 50 000mg/kg
(Examples: hydrides, Al dust)

(N.B. Class B1 consists of compounds of several metals while Class B2 lists a number of groups of organic compounds)

The nine technologies of origin include:

1. Treatment of metal surfaces;
2. Photochemical processes;
3. Wood impregnation;
4. Processes relating to textiles and printing;
5. Industrial reclamation of certain items, e.g. benzene, oil;
6. Cleaning of installations for storage and transport of oils and chemicals;
7. Petroleum refining and petrochemical production;
8. Certain primary production processes, e.g. pesticides, steel;
9. Laboratory chemical wastes - research and educational uses.

(N.B. Waste from any of these processes which does not contain substances listed in Classes A through D is not deemed chemical waste).

According to plans discussed in March, 1984, the list of proscribed materials is likely to be altered to include specific substances (probably to be taken from the German list or Austrian list). The purpose of this action will be to indicate that such substances are always chemical waste so that no chemical analyses will have to be performed by the generator.

NORWAY

The Pollution Control Act of 13 March, 1981 (H6) defines special waste regulations where issued on 10 April, 1984 which define the list of hazardous waste in Norway. The list was prepared cooperatively by the authorities and representatives of industry. A seventeen types of waste were selected; this list includes:

1. Waste oil;
2. Other oily waste, e.g. from tank cleaning;
3. Spent oil emulsions;
4. Organic solvents
   4.1 Spent solvents containing halogens;
   4.2 Spent solvents, without halogens;
5. Wastes containing paint, glue, varnish and printing inks;

6. Distillation residues;

7. Tarry waste including coal tars;

8. Mercury or cadmium containing waste;

9. Metal containing waste with lead, copper, zinc, chrome, nickel, arsenic, selenium or barium;

10. Cyanide containing waste;

11. Pesticide wastes;
12. PCB;
13. Isocyanates;
14. Other organic waste;
15. Strong acids;
16. Strong bases;
17. Other inorganic waste.

Of these, the first eleven must always be declared to the authorities.

SWEDEN

Swedish legislation (SFS 1975: 346; SFS 1983: 720) defines hazardous waste to be included under one the following types:

1. Waste oil;
2. Waste solvent;
3. Paint, varnish and adhesives wastes;
4. Concentrated acid or alkaline wastes;
5. Wastes containing cadmium, copper, chromium, nickel, tin or zinc resulting from surface treatment processes;

6. Wastes containing silver or zinc resulting from the printing or photographic industries;

7. Wastes containing mercury;

8. Wastes containing cyanides;

9. Wastes containing polychlorbiphenyl (PCB);

10. Wastes containing biocides.

A guidance document concerning hazardous wastes (SNV PM 690) issued by the Swedish Environment Protection Board gives detailed examples for each of these classes.

UNITED KINGDOM

The Control of Pollution (Special Waste) Regulations of 1980 list thirty-one substances as being classed hazardous wastes if they are dangerous to life by virtue of a single dose of no more than five cubic centimeters being able to cause death or serious tissue damage if ingested by a child of 20 kg body weight or exposure to the substance for 15 minutes or less would be likely to cause serious damage to human tissue by inhalation, skin or eye contact. Listed substances include:

1. Acids and alkalis;
2. Antimony and antimony compounds;
3. Arsenic compounds;
4. Asbestos (all chemical forms);
5. Barium compounds;
6. Beryllium and beryllium compounds;
7. Biocides and phytopharmaceutical substances;
8. Boron compounds;
9. Cadmium and cadmium compounds;
10. Copper compounds;
11. Heterocyclic organic compounds containing oxygen, nitrogen or sulphur;
12. Hexavalent chromium compounds;
13. Hydrocarbons and their oxygen, nitrogen and sulphur compounds;
14. Inorganic cyanides;
15. Inorganic halogen-containing compounds;
16. Inorganic sulphur-containing compounds;
17. Laboratory chemicals;
18. Lead compounds;
19. Mercury compounds;
20. Nickel and nickel compounds;
21. Organic halogen compounds, excluding inert polymeric materials;
22. Peroxides, chlorates, perchlorates and azides;
23. Pharmaceutical and veterinary compounds;
24. Phosphorus and its compounds;
25. Selenium and selenium compounds;
26. Silver compounds;
27. Tarry materials from refining and tar residues from distilling;
28. Tellurium and tellurium compounds;
29. Thallium and thallium compounds;
30. Vanadium compounds;
31. Zinc compounds.

GENERIC ITEMS LISTED BY THE UN RECOMMENDED
CLASSIFICATION SYSTEM (UN-RCS)

A.II.1. The UN Committee of Experts concluded that practical considerations prohibit the listing of all dangerous substances in specific terms. Therefore, the UN-RCS includes many generic or "not otherwise specified" (N.O.S) items. A list of most of these general items is included in this Annex.

A.II.2. Table of Generic Items in the UN-RCS

Identifier	UN-RCS Code (12)
Samples, explosive	0190
Articles, explosive (N.O.S.)	0349-0356
Substances, explosive (N.O.S.)	0357-0359
Ammonia solutions	1005
Driers, paint or varnish, liquid	1168
Extracts, aromatic, liquid (N.O.S.)	1169
Formaldehyde solutions, inflammable	1198
Ink, printers, inflammable	1210
Paints, enamels, lacquers, stains, shellac,) varnish, polishes, fillers (liquid), lacquer base or thinners, etc. (not including nitrocellulose)	1263
Perfumery products with inflammable solvents	1266
Petroleum distillates (N.O.S.)	1268
Tinctures, medical	1293
Coal tar distillates (inflammable)	1136
Ketones (liquid) N.D.S.	1224
Wood preservatives, liquid	1306
Inflammable solids (N.O.S.)	1325
Cotton waste, oily	1364
Driers, paint or varnish, solid (N.O.S.)	1371
Fibres, animal or vegetable N.O.S.) with animal or vegetable oil	1372
Fibres or fabrics, animal or vegetable (N.O.S.) with animal or vegetable oil	1373
Fuel pyrophoric (N.O.S.)	1375
Pyprophoric metals (N.O.S.) or pyrophoric alloys (N.O.S.)	1383
Wool waste, wet	1387
Alkali metal amalgams (N.O.S.)	1389
Alkali metal amides (N.O.S.)	1390
Alkali metal dispersions (N.O.S.) or Alkali earth metal dispersions (N.O.S.)	1391
Coal tar distillates (inflammable)	1136
Ketones, liquid (N.O.S.)	1224
Alkaline earth metal amalgams (N.O.S.)	1392

Alkaline earth metal alloys (N.O.S.)	1393
Hydrides, metal (N.O.S.)	1409
Bromates, inorganic (N.O.S.)	1450
Chlorates, inorganic (N.O.S.)	1461
Chlorites, inorganic (N.O.S.)	1462
Oxidizing substances (N.O.S.)	1479
Perchlorates, inorganic (N.O.S.)	1481
Permanganates, inorganic (N.O.S.) except ammonium permanganate	1482
Alkaloids (N.O.S.) or alkaloid salts (N.O.S.)toxic	1544
Antimony compounds, inorganic (N.O.S.)	1549
Beryllium compounds (N.O.S.)	1566
Chloropicrin mixtrues (N.O.S.)	1583
Disinfectants (N.O.S.) toxic	1601
Dyes (N.O.S.) or dye intermediates (N.O.S.) toxic	1602
Halogenated irritating liquids (N.O.S.)	1610
Nicotine compounds (N.O.S.) or nicotine preparations (N.O.S.)	1655
Rodenticides (N.O.S.)	1681
Thallium compounds (N.O.S.)	1707
Caustic alkali liquids (N.O.S.)	1719
Bifluorides (N.O.S.)	1759
Corrosive liquids (N.O.S.)	1760
Hypochlorite solutions (with more than 5% available chlorine)	1791
Medicines (N.O.S.)	1851
Rags, oily	1856
Textile waste, wet (N.O.S.)	1857
Resin solution, inflammable	1866
Disinfectants, corrosive, liquid (N.O.S.)	1903
Sludge acid	1906
Cyanide solutions	1935
Alcohols, toxic (N.O.S.)	1986
Alcohols (N.O.S.)	1987
Aldehydes, toxic (N.O.S.)	1988
Aldehydes (N.O.S.)	1989
Inflammable liquids, toxic (N.O.S.)	1992
Inflammable liquids (N.O.S.))	1994
Tars, liquid (including road) asphalt and oils, bitumen) and cut-backs)	1999
Metal alkyls (N.O.S.)	2003
Plastics, nitrocellulose based, spontaneously combustible (N.O.S.)	2006
Mercury compounds, liquid (N.O.S.)	2024
Mercury compounds, solid (N.O.S.)	2025
Phenylmercuric compounds (N.O.S.)	2026
Isocyanates, (N.O.S.)	2296; 2207
Blue asbestos	2212
Lead compounds, soluble (N.O.S.)	2291
Organic peroxides, samples (N.O.S.)	2255
hydrocarbons (N.O.S.)	2319

Alkyl phenols (N.O.S.) including
C$_2$-C$_8$ homologues)   2430
Cadmium compounds except cadmium selenide
and cadmium sulphide   2570
Pesticides, solid, toxic (N.O.S.)   2588
Alkylamines (N.O.S.) or   )
polyalkylamines (N.O.S.)   )
(inflammable, corrosive)   )   2733-2735
Chloroformates (N.O.S. - flashpoint
not less than 23°C   2742
Organic peroxides, mixtures   2756
Organotin compounds (N.O.S.)   2788
Substances which, in contact with   )
water, emit inflammable   )
gases (N.O.S.)   )   2813
Infectious substances, human (N.O.S.   2814
Dyes (N.O.S.) or dye intermediates (N.O.S.),
corrosive   2801
Poisonous liquids (N.O.S.)   2810
Poisonous solids (N.O.S.)   2811
Pyrophoric liquids (N.O.S.)   2845
Pyrophoic solids (N.O.S.)   2846
Infectious substances, non-human (N.O.S.)   2900
Pesticides, liquid, toxic   2902; 2993
Corrosive liquids, inflammable (N.O.S.)   2920
Corrosive solids, inflammable (N.O.S.)   2921
Corrosive liquids, poisonous (N.O.S.)   2922
Corrosive solids, poisonous (N.O.S.)   2923
Inflammable liquids, corrosive (N.O.S.)   2924
Inflammable solids, corrosive (N.O.S.)   2925
Inflammable solids, poisonous (N.O.S.)   2926
Poisonous liquids, corrosive (N.O.S.)   2927
Poisonous solids, corrosive (N.O.S.)   2928
Poisonous liquids, inflammable (N.O.S.)   2929
Poisonous solids, inflammable (N.O.S.)   2930
Pesticides, liquid, inflammable,   )
toxic (N.O.S.)   )
(flashpoint less than 23°C)   )   3021

NATIONAL MONITORING SYSTEMS FOR HAZARDOUS WASTE

F. Van Veen
Consultant to the Environment Directorate,
OECD, Paris

SUMMARY

Based on information supplied by several OECD countries, a brief description of existing monitoring systems is presented. Notification and transport document requirements are emphasized. Results indicate that each nation requires notification and transport documents which differ from one another. The Directive of the European Communities concerning transfrontier shipments is likely to aid in reducing transport difficulties which may arise as a result of several systems being in force simultaneously. A very brief survey of international rules governing transport of dangerous goods is also provided.

INTRODUCTION

Imports and exports of hazardous wastes occur on a large scale especially in Europe. Differences in legislation, differences in disposal prices and/or availability of appropriate treatment, storage and disposal facilities are key factors which influence the volume of transfrontier movement of hazardous wastes.

Such traffic in wastes does not only occur between industrialized countries, e.g. EEC and OECD members. Hazardous wastes may also flow to less developed countries (LDC's). Such countries may not have appropriate legislation in place meant to protect the local environment especially with respect to contamination of soil and groundwater.

Accurate information concerning hazardous waste generation, transport and mode of disposal is difficult to obtain. Nevertheless, detailed information concerning types and quantities of wastes which cross national frontiers is necessary in order to effectively control such traffic.

No legally constituted system exists which provides information and data with respect to transfrontier movement of hazardous wastes. However, several countries, as required in their internal hazardous waste legislation, do collect certain data concerning traffic in hazardous wastes. Data obtained by means of these national information (notification) systems may well be of great importance to efforts aimed at harmonizing rules relating to imports and exports of hazardous wastes.

In order to appropriately monitor export-import activities, transmission of information between exporting and importing countries must be facilitated, definitions of hazardous wastes must be placed on a common basis, and control procedures at the border-stations must be improved.

The primary requirement is implementation of adequate hazardous waste legislation including appropriate import-export provisions. Then "cradle-to-grave" control can be maintained if necessary. Finally, licensed disposal facilities must be assured in order for the control system to function properly.

MONITORING SYSTEMS IN SEVERAL OECD COUNTRIES

This section, briefly summarises the notification- and trip-ticket systems used in several countries. Details for each country are contained in Annexes 1 to 11. Annex 12 contains a survey of international agreements and rules governing transport of dangerous goods.

1. France

The principal laws are the Law of 15th July 1975 (Elimination of Waste and Recovery of Materials) and the Law of 19th July 1976 (Classification of Installations for Environmental Protection).

The Law of 1975 places primary obligation for appropriate hazardous waste management practices onto the generators. The Law applies to waste in general as well as to certain special types of wastes. The Decree of 19th August 1977 lists these special (toxic) wastes for which the authorities can require generators, transporters and disposers to provide complete information.

The Law of 1976 regulates, in particular, the construction and operation of waste treatment, storage and disposal sites. These sites are subject to a prefectural authorization which stipulates the technical requirements of their operation. Such requirements are indicated in technical instructions issued at national level :

-- instruction of 22nd January 1980 concerning landfilling of industrial waste ;

-- instruction of 21st March 1983 concerning incineration of industrial waste.

At present a number of waste generators, collectors and disposers are required to report to the prefectural authorities, periodically, data regarding the nature, quantities and destination of the special waste they generate or accept for treatment. For the classified installations (generators and disposers) this requirement is usually embodied in the decree authorizing the installation. This reporting system will soon be standardised; data will have to be provided every three months to the administrative authority (Direction Régionale de l'Industrie et de la Recherche) by means of standard notification forms (see Annex 1). The new system will be founded upon computerized manipulation and storage of the data.

There is currently no manifest sytem specifically applicable to hazardous waste transport; standard transport documents as required by the regulations governing the transport of dangerous goods are utilized. However, requirements for a specific manifest to accompany the waste throughout its entire journey will soon be implemented.

With respect to the import of waste, the Regulation of 5th July 1983 requires the importer to furnish a preliminary declaration of intent jointly signed by the generator, the transporter and the importer. This declaration must be transmitted to the authority responsible for the control of treatment facilities. This authority must ascertain and affirm that the waste in question can be disposed satisfactorily at the proposed facility; without such affirmation import of the waste can be prohibited. If affirmation is obtained, the import declaration must be presented by the waste transporter to Customs officials when the waste actually crosses the border. A blanket affirmation allowing regular shipments of the same waste to a given facility may be implemented.

In cases of transit of hazardous wastes through France a similar procedure is prescribed by the Regulation of 5th July 1983.

The Regulation of 5th July 1983 also lists the categories of waste for which a preliminary import declaration is required, and indicates the information to be provided in the declaration.

2. Germany

The Federal Waste Disposal Act regulates hazardous waste management, (7th June 1972, amended on 21st June 1976 and implemented on 1st January 1977).

The order relating to "hazardous waste" of 24th May 1977 lists specific wastes relegated into this category.

Wastes may only be treated, stored, deposited, collected and transported by licensed persons.

The Administrative Order regarding Notification of 29th July 1974, amended by the Order of 2nd June 1978, provides rules for manifest procedures, record-keeping and other control measures. Manifests and appropriate record-keeping are mandatory for the generator of hazardous wastes. Record-books which contain copies of manifests must be maintained by generators, transporters and disposers. Additional copies of the manifest must be sent to the responsible state authorities at the beginning and the end of the disposal process. Such notification provides for "cradle-to-grave" control. A description of the manifest is given in Annex 2.

Import of wastes is regulated by paragraph 13 of the Waste Disposal Act in conjunction with paragraph 1 of the Administrative Order on Import of Waste. It is agreed practice that for the import of waste, a written declaration is required from the authorities of the exporting country that the waste cannot be treated in the exporting country. A full analysis of the waste is required for imports.

At present, export only can take place in connection with a licence system. For transit no special rules exist.

Control is accomplished at the State (Land) level. Authorities at this level required at least 3 years in order to become fully familiar with the manifest system and the processing of the data. The current estimate is that about 90 per cent of the hazardous waste generated is covered by this notification system. The present number of manifests is estimated to be 1-1,3 million.

A loophole in the monitoring system is that certain persons may take improper advantage of differences in the respective definitions of hazardous waste and secondary (scrap) feed stock materials. In the case of import-export, such selective definition of the material being transported can also occur. A proposed amendment to the waste disposal act, giving power to also control certain materials that are intented for recycling is currently discussed by parliament.

3. The Netherlands

The Chemical Waste Act of 20th April 1976 is the main legal instrument allowing for the management of hazardous wastes. Implementation of this act took place on 1st August 1979.

Provisions of the Act require a licensing system for treatment, storage and disposal of hazardous wastes.

The act itself provides the legal framework. Under its auspices, several Decrees and Orders have been published: e.g. the Materials and Processes Decree (definition of chemical waste) and the Chemical Waste Notification Decree.

The implementation of the Act includes the introduction of a notification system and means for processing of the information received. Currently, more than 50.000 notifications per year are processed at the central government headquarters; the notification system is centralised within the Ministry of Housing, Physical Planning and Environment. Computers are used to store the information received from each person/firm. The notification system (a "cradle-to-grave" system) enables one to decide whether the waste is properly disposed. The system also provides an opportunity to trace those generators suspected of failing to notify. At present, information is available for more than 70 per cent of the chemical waste flow (this percentage is an estimate based on scientific inventories).

The Chemical Waste Act only applies to waste which is disposed of off-site. On-site land disposal is prohibited in the Netherlands.

In Annex 3 a description concerning aspects of the notification procedure is given; manifests for hazardous waste do not exist. For transport activities, normal transport-documents have to suffice.

Export is, in principle, free since only notification has to take place. Import of hazardous waste is forbiden, except in cases where such waste is to be disposed by a permitholder. At the border, the normal (import-export) controls are applied. Because actual control is not possible, customs officials occasionally contact the Ministry of Environment for further data and interpretation of the information given on the transport manifest. At present all border transit points can be used for waste. In future, places for transboundary transit may possibly be restricted.

Transit-rules are applied according to the normal commodity transit regulations.

Record-keeping by the disposer is practiced. In principle such records are unnecessary, since all information is centralized and available at the Ministry.

Difficulties associated with transboundary activities result from differences in definitions of hazardous wastes between countries and from lack of information provided by the exporting- to the importing country.

4. Sweden

The Ordinance on Environmentally Hazardous Waste of 1st January 1976 is the main legal instrument governing the management of hazardous wastes.

There is currently no notification- and/or manifest-system in operation. The requirement of the ordinance is that generators of hazardous waste must make a declaration to municipal authorities on an annual basis. This declaration must include information concerning nature of the waste as well as the method of disposal. The present system is unsatisfactory from an operational point of view. Therefore, the Swedish government is preparing a more comprehensive system. Details of this proposed system are not available now, but will be published in the near future. Regarding the notification system one could perhaps speak of a system on the community level. See Annex 4.

No import rules exist; thus, no restrictions on import exist. Companies which transport and manage hazardous wastes must have the necessary permits whether or not the wastes are of Swedish origin.

A permit issued by the Environmental Protection Board is necessary to export hazardous waste unless the export is carried out by SAKAB, the Swedish national disposal contractor. An export licence is primarily granted for those wastes which cannot be appropriately treated in Sweden. While a single export licence may be granted for several batches of waste, a new licence is usually required for individual batches.

No special transit rules exist. Transit usually does not occur, but in the case of transit, normal transport rules are applied.

On-site treatment, storage and disposal of hazardous waste is not regulated by the the Ordinance on Hazardous Waste. Licensing and surveillance are carried out in the same manner as for other industrial activities producing similar substances. This mechanism also applies to intermediate storage in cases where no specific regulations exist.

5. Switzerland

Federal legislation currently in force does not provide the necessary legal basis for effective control of waste flows.

The situation will change soon: a law dealing with Environmental Protection was adopted by Parliament on 7th October 1983. Based on this law, the Federal Office for Environmental Protection has prepared a Draft Ordinance for the Control of Hazardous Waste.

The proposed Ordinance calls for a notification requirement combined with a manifest system similar to the one which is used in the Federal Republic of Germany. A description of the Swiss version is given in Annex 5.

The Federal Council or the cantons can order that records be kept concerning hazardous waste, as well as on types and quantities of raw materials and other products. Whether record keeping will be required on a regular basis is not currently known.

The proposed Ordinance will only regulate off-site treatment. Hence, notification is not necessary for on-site treatment. In exceptional cases, hazardous waste generators may be obliged to maintain records for wastes treated on-site, but this requirement is beyond the scope of the proposed Ordinance.

Annex 5 provides a description of a draft notification form.

All information provided by the notification system will be collected at the Federal level at a single "information collection center". Statistics based on these data will be published on an annual basis. The manifest must be completed prior to shipment; copies are sent to the government after shipment.

Import of hazardous waste can only occur if a Swiss disposer who has obtained a permit for accepting and importing waste is ready to accept the waste in question.

Flaws in the import-export system will occur if hazardous waste is not declared as such or if the declaration is incorrect. Such flaws cannot be eliminated by the national legislation of the importing country.

6. United Kingdom

The Control of the Pollution Act 1974 (COPA) provides the statutory framework for waste management. COPA requires that industrial waste be disposed at a licensed site. This license system is fundamental to the UK management programme.

In addition, documentary control over the movement of the most toxic and dangerous wastes is maintained. The control system and the wastes which fall under its jurisdiction, are defined in the Control of Pollution (Special Waste) Regulation of 1980.

Administrative responsibility for appropriate control of hazardous waste lies with the waste disposal authorities (WDA) of the local Councils. WDAs are responsible for the issue or site licenses, for monitoring and surveillance and for prosecution of alledged offenders.

Under COPA, WDAs are also required to survey wastes originating in their respective areas and to prepare a waste disposal plan.

Currently 500 disposal facilities are licensed: 200 on-site, 200 operated by commercial disposal contractors and 100 by the WDAs. Individual facilities are often small in terms of licensed capacity. More than 95 per cent of the total is handled by the commercial contractors.

• The so-called "consigment-note" (manifest) which is a pre-notification system meant to help achieve "cradle-to-grave" control is described in Annex 6. This document accompanies all transport of the waste; additionally, the Regulations of 1980 list record-keeping requirements.

Import of hazardous wastes is not prohibited. Under COPA, the importer is considered to be the generator. Thus, the importer must complete the consignment note. The result is that cases of import activity are reported to the WDA.

Export of hazardous wastes is not prohibited. Under COPA, the exporter is considered to be the disposer. Hence, the manifest must be completed in the usual fashion; thus, in cases of export of hazardous waste the WDAs are informed.

The Licensing Regulations do not currently cover wastes in temporary storage but an amending Regulation planned to be introduced in 1984 will, with certain exemptions, cover such wastes.

Record-maintenance and retention is necessary for generators and transporters (both 2 years) and for the disposer (indefinitely).

7. United States

The Resource Conservation and Recovery Act (RCRA) as amended is the principal law governing control and management of hazardous wastes. The regulatory system, associated with RCRA, has five major elements:

-- federal classification of hazardous waste

-- cradle-to-grave manifest

-- federal standards for safeguards to be applied by generators, transporters and facilities which treat, store or dispose of hazardous waste

-- enforcement of federal standards for facilities through a permitting (licensing) programme

-- authorizations of individual State programmes.

Before shipping a hazardous waste off-site, the generator must prepare a manifest identifying the transporter and designating the treatment, storage or disposal facility. A uniform manifest, to be used throughout the United-States, was promulgalted in the Federal Register in early 1984. This manifest will also be used in case of international transport of hazardous waste involving the U.S. portion of the shipment.

The manifest procedure is as follows (see Annex 7):

The generator completes the manifest and retains a copy. Other copies go to the transporter and the designated disposer. Each transporter and disposer keeps a copy and the disposer, after accepting the waste, returns a copy to the generator.

Record-keeping is required of the generator, transporter and disposer; records must be retained for a period of 3 years.

EPA does not require notice of shipments or copies of the manifests. Thus no information is available from this source on existing waste flows. Information on waste flows is collected biannually.

Currently EPA is informed only in cases of accidents and spills or where the generator does not receive a return copy of the completed manifest. By monitoring these cases, EPA found that manifests are apparently not properly completed in all instances.

Many states have manifest systems in operation. Several State Agencies do require reports of movements of hazardous wastes.

Export-import is not prohibited by RCRA. The export-import of PCB waste is prohibited by the Toxic Substances Control Act. Export of dioxin wastes can be prohibited; the intent to export dioxin or indeed to ship it within the country must be reported to the U.S. Environmental Protection Agency (EPA) 60 days in advance.

In the case of export of hazardous wastes, EPA must be notified four weeks prior to the shipment. EPA notifies the importing country and may provide, if requested, certain technical assistance. Customs control is not special; normal procedures for hazardous materials are followed.

89

A recent proposal to amend RCRA was placed before the U.S. Congress (July, 1983). This amendment would govern export of hazardous waste from the U.S. by requiring -- two years after enactment -- that a prospective exporter provide EPA with:

1. Name and address of exporter

2. Types and estimated quantities of hazardous waste to be exported

3. Estimated frequency of export activity and the period of time over which waste will be exported

4. Ports of entry

5. A description of the manner in which such hazardous waste will be transported

6. Treatment, storage or disposal method to be used in the receiving country

7. Name and address of ultimate treatment, storage or disposal facility.

In turn, EPA would notify the destination country and request prior written consent of the receiving country. EPA would also advise that export is prohibited unless such consent is received. A copy of the applicable U.S. regulations governing the treatment storage and disposal of the waste if it remained in the U.S. would also be forwarded. Once, consent or objection were received, the prospective exporter would be notified. All exporters of hazardous waste would be required to submit an annual report to EPA summarizing export activity.

EPA is also authorized to enter into bilateral agreements with receiving countries which may establish alternate notification procedures and eliminate prior written consent for each export of hazardous waste.

8. Denmark

The Act on the Disposal of Oil and Chemical Waste of May 1972, provides the statutory framework for waste management in Denmark.

A number of statutory orders have been issued in pursuance of the Act, i.e.:

-- Statutory Order N° 121 of March 17, 1976, on Chemical Waste, as amended by Statutory Order N° 323 of July 3, 1980, and

-- Statutory Order N° 410 of July 27, 1977, on Oil Waste.

In accordance with the provisions of these texts, hazardous waste is defined as:

Chemical Waste: Waste to which the Annex of Statutory Order on Chemical Waste applies, and types of waste with similar properties (for instance corrosive, toxic or flammable).

Oil Waste: Products containing mineral oil, or products containing
                    synthetic oil, which are no longer intended for use in their
                    present state.

The rules applying to waste intended for reuse, and waste intended for
destruction, are the same.

As a general rule undertakings handling hazardous waste shall deliver
the waste to a site prescribed by the municipal authorities.

If industrial undertakings wish to find ways of disposal of hazardous
waste other than storage in the municipal waste collection station, a licence
shall be granted by the municipal authority, no matter whether the waste is
designated for export or disposal in Denmark. The municipal authority shall
issue the licence if the undertaking is able to prove that the waste will be
safely transported and disposed of.

Import of chemical waste to Denmark shall be notified by the importer
to the municipal authority at least one month before the waste is imported.

Undertakings handling hazardous waste shall, as specified in detailed
regulations, observe the rules concerning approval and supervision laid down
in Act on Environmental Protection.

Transport of hazardous waste is regulated in legislation laid down by
the Ministry of Justice, implementing the provisions of international
conventions on the transportation of dangerous goods (ADR, RID, and IMO).
Detailed rules have been laid down for instance on packaging of various types
of waste, and on the approval of motor vehicles used for road transport of
hazardous waste.

Specific monitoring rules as to import, transit and export do not exist
in Denmark at present.

9. Canada

The transportation of Dangerous Goods Act of 1980 and accompanying
draft regulations (to be enacted in early 1984) respecting the handling,
offering for transport and transport of dangerous goods will together comprise
the main statutory instrument for the management of hazardous waste in Canada
at the federal level.

The provinces are responsible for the siting, design, approval,
licensing, monitoring and surveillance of treatment disposal facilities and
the administration and enforcement of provincial legislation.

In Annex 8 a description is given of the coming Transportation of
Dangerous Goods Act as far as hazardous waste is concerned. In this annex
also the proposed manifest is given. Discussions with the provincial
regulatory authorities regarding the possibility of adopting it as a uniform
manifest are in the final stage.

Discussions with US-EPA are beginning as to the appropriate pre-notification process and procedures for import-export between the two countries.

10. Finland

The principal law regulating the waste management in general as well as hazardous waste is the Waste Management Act from 1978. According to the Act the management of hazardous waste is supervised by the authorities both at the site of generation and the site of treatment, using two ways of authorisation, i.e. a waste management plan for a real estate or a permit to treat hazardous waste. There is no manifest system in the domestic transport, so each shipment of hazardous waste cannot be controlled. The generator has to report the authorities the place to where the waste is going to be transported in case it is not a municipal treatment site. The place of treatment or disposal has to report the place of origine of the waste it treats. Changes in this respect must also be reported. The Waste Management Act allows transport of hazardous waste only to a municipal treatment site or to an approved other site. The export and import of hazardous waste must also be mentioned when applying for an authorisation. For each single importation and exportation a specific notification is still required. This system makes it possible for the authorities to be aware in advance of the main streams of hazardous waste. The control afterwards of single transportations is to some extent possible by checking the records held by the generator or the disposer.

Import and export of hazardous waste is to be notified to the Ministry of the Environment at least 30 days before the importation or exportation takes place. The Ministry may prohibit the importation or exportation of hazardous waste if there are proper reasons to suspect that the waste will not be transported or treated in a way that can be regarded as appropriate from the viewpoint of environmental protection or waste management.

In practive the Ministry of the Environment reports the authorities of the importing country that such an export has occurred and requires the exporter to show a document from the treatment place proving that the waste concerned has been received. The transit countries are at the moment not informed. Customs procedures are the same as for usual hazardous materials.

The monitoring system described above has been fully in force since october 1981, and is still under development. The limited resources in the administration also retards the implementation of provisions concerning the monitoring both in the domestic and transfrontier movements of hazardous wastes.

11. Austria

The Federal Government of Austria is responsible for control of wastes produced in trade and industry. In addition, by virtue of the Federal Act of 7th March 1979, the Federal Government has also been given special responsibilities with respect to waste oil treatment.

Annex 9 contains a description of the combined notification and manifest system for hazardous wastes.

The Austrian manifest is comparable with the trip-ticket of Germany (see Annex 2); only small differences exist. One major exception is that, for Austria, the transporter must notify waste whereabouts to the responsible governmental authority rather than the generator. On the other hand, a generator must provide advance notice of his basic waste position to the Austrian authorities.

12. Norway

The Act of 13th March 1981 Concerning Protection against Pollution and Concerning Waste was implemented on 1st October 1983. Pursuant to this act, Regulations on Delivery, Collection, Reception and Treatment/Disposal of Certain Types of Hazardous Waste were issues on 10th April 1984.

The regulations state that enterprises generating annually more than listed amounts of 11 given types of hazardous waste shall deliver the waste to authorized waste handlers (reception stations or treatment plants).

All enterprises dealing with collection, reception, treatment, exportation or importation of the given types of hazardous waste shall be authorized by the Environment Protection Authorities. The authorization states regulations on handling, treatment, journalizing etc. Enterprises dealing with transportation of hazardous waste do not need any authorization, provided they just transport the waste to a given receiver. If the transporter offers to "take care" of the waste, this is regarded as collection and authorization is compulsory.

For exportation of hazardous waste, a pre-notification to the environmental authorities in the receiving country is mandatory. An acceptance of the pre-notification may also be demanded before the exportation can take place.

By delivery of waste to a waste receiver, a waste-declaration must accompany the delivery. The declaration acts as a trip ticket. Further information on the system is given in Annex 10.

DIRECTIVE OF THE EEC CONCERNING TRANSBOUNDARY TRANSPORT OF HAZARDOUS WASTES

As a result of the adoption of the EEC Directive, European countries will have to change their legislation or to enlarge their jurisdictional control over hazardous wastes.

A brief description of the EEC directive concerning transfrontier shipment is as follows:

The main feature of the proposal as related to this report is the suggested notification and manifest system.

Article 3 states that if hazardous wastes (as defined by the EEC) are to be shipped to another country, the generator shall provide notice to the competent authorities of the country of destination, of dispatch of the shipment and of transit. Such notification shall be made on a standardized form which is given in Annex 11.

Article 4 states that the shipment may not be executed prior to receipt of acknowledgement of the notification by the authorities of the country of destination. A copy of this acknowledgement shall accompany the shipment.

Article 4 also says that objections against the transport may be in order. The grounds for such objections must be stated, however.

The Directive also lists rules for the consignment document (manifest). A description is given in Annex 11.

The goal of the directive is to improve the information available to the competent authorities and to better regulate export and import of hazardous wastes. (Note that the recent proposal in the U.S. to amend RCRA is very similar to the EEC Directive.)

DISCUSSION

Aspects of the monitoring systems of ten OECD countries are shown in Table 1. These aspects are discussed below:

For an effective monitoring system the first prerequisite is adequate legislation. Switzerland is reviewing its current legislation. Sweden and Germany are strengthening current law in the areas of notification and control. A wide variety of amendments to the U.S. Law have been proposed and are currently under review.

The degree of implementation differs from country-to-country and can not be quantified. For example, an estimate from the Netherlands is that about 80 per cent of the total amount of hazardous wastes is notified and treated according to the law. In other countries this percentage may be lower.

For effective control of hazardous waste a "cradle-to-grave" monitoring system seems to be desirable. In principle, such control is necessary for each shipment of waste. Control can be achieved through notification made by the generator and by the disposer at the time of acceptance of the waste transport. In Austria, notification must be accomplished by the transporter instead of the generator, hence, very few transporters will be licensed under the hazardous waste disposal act. However, not all OECD countries have such a "cradle-to-grave" monitoring system.

On-site treatment for most OECD countries is not regulated by existing hazardous wastes legislation. This does not apply for the FRG and the USA, where licensing is mandatory, as well as for the Netherlands, where on-site land disposal is prohibited. However, in exceptional cases, notification can be required. Control of on-site treatment, storage and disposal activities is then possible in the context of other legislation such as air- and water pollution acts and/or public nuisance acts. Control of the quality and quantity of the treated wastes does not appear possible at present.

Most countries have a reporting system in case of transport of hazardous wastes. Only the United States of America has a manifest system where notification to the Authorities (EPA) is not required although some states do require notification. In the USA a biannual report is required but not on a shipment by shipment basis.

Elsewhere, reporting may be required for each waste transport (U.K., Switzerland and Germany), monthly (The Netherlands), on a quaterly basis (France), or on an annual basis (Sweden). In The Netherlands, a waste type being shipped for the first time must be described in a very detailed way. When this first waste transport has taken place, subsequent transports (of the same wastes to the same disposer) can be reported in less detail.

Reporting can occur in advance or not. The rule seems to be that where each waste shipment must be reported, notification takes place in advance.

Not all countries have a hazardous waste manifest system. Three countries (U.K., Switzerland and Germany) have a combined reporting and manifest system; France will soon apply a manifest system in addition to the quaterly notification system. The USA have recently adopted a uniform manifest system. Other countries make use of the normal transport documents. On these documents, there should be some indication that a hazardous waste is undergoing transport. However, in such cases no information is available on aspects of the waste which might be important for customs and control agents.

Record-keeping is not always required. Furthermore, if information from required reports is processed at a central level, record-keeping by others might not be necessary.

In general, import-export of hazardous wastes can only be prohibited if the intended disposal activities are not licensed, if proper notification, if required, is not reported and/or if the manifest and transport-documents are improperly prepared.

Some countries (e.g. the FRG) will only accept waste for import if it is accompanied by a written declaration from the authorities of the exporting country that the waste cannot be treated there.

Notification in the case of import-export of hazardous waste is not always necessary. Some countries (Sweden, USA) are not informed when wastes are imported. Thus control is made more difficult as shown by the Seveso waste affair.

Some countries (U.K., France, Switzerland in the future, and the Netherlands) seem to be well-informed about what is occurring at their borders with respect to hazardous wastes. Some countries (e.g. the USA) transmit information to the importing country that a waste will be exported. These

Table 1

ASPECTS OF MONITORING SYSTEMS

	UK	Switz	Swed.	FRG	USA	NL	DK	CAN	F (1)	F (2)	N
Hazardous Waste Law:											
Date of implementation	'72 '80	-	'76	'77	'80	'79	'76	-	'77	-	'83
On Federal, State, Municipal level?	S	F/S	F/M	S	F/S	F	M/S	F/S	F/S	F/S/M	F/S
Cradle to grave control?	+	+	-	+	+	+	-	+	-	+	+
Notification:											
For on-site treatment?	-	-	-	+	-	-	+/-	-	-/(+)	+	-
For off-site treatment?	+	+	-	+	-	+/-	+/-	-	+	+	+
Per waste transport?	+	+	-	+	-	+/-	-	-	+	+	-
For regular shipments?	-	-	-	-	-	+	+/-	-	-/(+)	+	-
Is notif. in advance?	+	-	+	+	+	-	+/-	-	-	-	-
Trip-ticket:											
Exist a special waste Trip-ticket?	+	+	+	+	+	-	+	+	+/(-)	+	+
Are notification and trip-ticket combined?	+	+	-	+	-	-	-	+	+	+	+
Record-keeping:											
Is record keeping obliged?	+	+	+	+	+	+/-	-	+	+	+	+
Export-import:											
Is notification needed for export?	+	+	+	+	+	+	+/-	-	-	+/(-)	+
Is notification needed for import?	+	+	-	+	-	+/-	+	-	+	+	-
Takes special control place (sampling + analyse) at border?	-	-	-	-	-	-	-	-	+/-	+/(-)	-

(1) present system.
(2) proposed system.

transmissions can only be accomplished when notification takes place in advance. Effective monitoring in the importing country would seem to be promoted by such information exchange.

Control of hazardous waste at border points is comparable to control of normal commodities and chemicals. Some analytical means of control may be possible, but such action is costly and does not assure complete control. If border-stations for hazardous wastes transboundary shipments are restricted, customs officials can be trained and the stations can be equiped better for analytical controls and sampling.

Exceptions to reporting requirements exist in several countries. Small waste quantities may be excepted (Switzerland, U.S.A. and The Netherlands), wastes with high economic value (Switzerland), wastes for recycling and recovery (USA) and central collection-stations (Switzerland and The Netherlands). Note that the U.S. appears likely to reduce the small generator exception from 1000 kg per month to 100 kg per month.

Flaws in the regulations of import-export activities exist. The main causes of such flaws are failure to transmit information to importing countries and individual differences in legislation and in definitions of hazardous wastes. A special case is the monitoring system of the U.K., where the importer is taken to be the generator and where the generator must report to the WDA. A potential difficulty with this system is that in the case of an illegal activity, the original waste generator from abroad can only be prosecuted under his own country's disposal law.

In Table 2 aspects of the documents used in the United Kingdom, Switzerland, Germany, The Netherlands, the United States and France are given.

In the U.K., Germany and Netherlands experience is available based on the documents. Processing and sorting the resultant information in such a way that "cradle-to-grave control" can be maintained has not been fully achieved. The best means to this end seems to be a combined manifest and notification-document. Inventories can be published, however. The training period for staff of authorities can, according to statements from Germany, be several years.

The information on the document may be relatively simple (Germany) or rather detailed (Netherlands).

A simple document requires information concerning:
- generator
- type (waste-stream/number) and quantity of the waste
- properties of the waste
- transporter
- disposer
Additional information is sometimes required including:
- composition of the waste
- properties of the waste in relation to danger and risk
- date of transport
- method of disposal

Information can be simplified by making use of code-numbers. These numbers can be used for:

- generator
- type of waste
- transporter
- disposer
- method of disposal

These numbers can be assigned by the competent authorities. A code-number for the waste type is more or less usual in several countries (also called waste stream number).

Table 2

INFORMATION TO BE FILLED IN ON THE DOCUMENTS (1)

	U.K.	Switz.(2)	FRG	NL	USA	F	N
Information is given on:							
Generator	+	+	+	+	+	+	+
Contact person	+	-	-	+	-	+	+
Type of waste	+	+	+	+	+	+	+
Wastestreamnumber (code)	-	+	+	+	+	+	-
Quantity	+	+	+	+	+	+	+
Composition	+	-	-(3)	+	-	+	+
Consistency	+	+	+	+	-	+	+
Properties	+	+	-	+	-	+	+
Production-process	+	-	-	+	-	+	+
Transporter	+	-	+	+	+	+	-
Transportvehicle (reg. number)	+	-	+	-	-	+	-
Packaging	+	-	-	+	-	+	-
Date of transport	+	-	+	-	-	+	-
Disposer	+	+	+	+	+	+	+
Way of disposal	-	-	-	+	-	+	+

* In the United-Kingdom, Switzerland, Norway and Germany the notification record and the trip-ticket are combined; the Netherlands have no trip-ticket and the United-States have no notification.

** Document in draft form.

*** Composition can be included by using the nomenclature of the waste-list according to § 2 para. 2 of Waste Disposal Act.

- Monitoring systems for hazardous waste differ from country to country.

- Most countries have a notification-system either per waste transport or quarterly.

- Manifest documents are less common, but where such a system is in place, the notification and manifest document is combined except in the USA where reporting to the Federal Authorities is required only biannually. The European Commission has proposed both a notification and a manifest document, but separate documents are to be completed.

- Cradle-to-grave control is currently impossible in some OECD countries. However, where each transport action must be reported in advance, such control can be obtained.

- Notification of intent to import and export is not required by some countries. The exporting country does not always inform the importing country. Thus, the procedures for import-export of hazardous waste differ from country to country.

- Special control at the border is extremely difficult to arrange. In practice, control is usually comparable with the control of hazardous commodities or dangerous goods.

- In some cases, small quantities of waste are exempt from reporting requirements.

- Flaws occur in import-export activities involving hazardous wastes, especially in cases of international differences in definitions of hazardous wastes and -- most important -- as a result of not informing the importing country that a hazardous waste is arriving at its frontier.

- All the official notification- and manifest documents differ depending on the country. In general, the manifest requires the same information about generator, type and quantity of waste, transporter, transport-vehicle, type of packaging and disposer. But, information about composition and properties of the waste, date of transport and way of disposal is not always necessary. Also the code-numbers for the type of waste differ from country to country.

- No special regulations exist for transit-activities, except the normal transit-procedures for hazardous commodities.

- The European Commission has proposed a mechanism for transmission of information for transit and import shipments to prospective transit nations and importers. A proposal before the U.S. Congress is somewhat similar.

NOTES AND REFERENCES

1. Hazardous Waste Legislation in OECD Countries, OECD, Paris 1983.

2. Industrial Hazardous Waste Management, Industry and Environment (Special issue), No. 4 1983.

Annex 1

FRANCE
(proposed system)

Notification Systems (every three months)

Three reporting forms exist:

-- Form A for the generator
-- Form B for the collector/transporter
-- Form C for the disposer

On the forms information must be given concerning the parties of interest (generator, transporter and disposer), the waste (quantity, code, place of origin), the means of transport, packaging and finally about the method of disposal.

The forms must be sent to the responsible authorities four times per year.

Trip-Ticket System

A trip-ticket (bordereau de suivi) must be completed by the gneerator when he transfers his waste to another person. This trip-ticket must indicate the origin, the characteristics and the destination of the waste, the arrangements made for their collection, transport, disposal or storage, as well as the identification of the firms involved in these operations.

The trip-ticket accompanies the waste until final disposal. The ticket must be signed successively by the generator, the collector, the transporter, the disposer and, if applicable, the owner of the transit centre. Each signatory keeps one copy of the trip-ticket.

The disposer must transmit to the generator, within two weeks, a signed copy of the trip-ticket indicating that the waste has been either accepted or rejected. In case of rejection, the disposer warns the responsible authority.

The disposer must certify to the generator that the waste has been eliminated. If the generator does not receive from the disposer, within two weeks, the signed copy of the trip-ticket certifying that the waste has been accepted, he must alert the responsible authority.

Notification forms

DECLARATIONS DE PRODUCTION DE DECHETS INDUSTRIELS

ENTREPRISE PRODUCTRICE		PERIODE
Raison sociale :　　　　　N° SIRET :		TRIMESTRE :
Lieu de production :　　　Activité :		ANNEE :
Tél. :　　Nom du responsable :　　Visa :		

DESIGNATION DU DECHET	CODE NOMENCLA-TURE R | A	ORIGINE Atelier de pro-duction/process	COLLECTEUR (1)	Quantités en tonnes	DESTINATION (1)	
					ENTREPRISE DESTI-NATAIRE (1)	Mode de traitement interne (

(1) Raison sociale et localisation

(2) Cette colonne ne doit être remplie que si les déchets sont éliminés au sein de l'entreprise productrice. On utilise le code suivant : D : décharge - PC : traitement physico-chimique - S : station d'épuration - E : égoût - N : rejet en milieu naturel

DECLARATION DE COLLECTE DE DECHETS INDUSTRIELS

ENTREPRISE DE COLLECTE		PERIODE
Raison sociale :　　　　　N° SIRET :		TRIMESTRE :
Adresse :		ANNEE :
Tél. :　　Nom du responsable :　　Visa :		

PRODUCTEUR DU DECHET (1)	DESIGNATION DU DECHET	CODE NOMENCLA-TURE R | A	Qté en tonnes	Mode de transport	CONDITION-NEMENT DU DECHET (2)	ENTREPRISE DESTINATAIRE (1

(1) Raison sociale et localisation

(2) On utilise le code suivant : V : Vrac - F : Fûts - C : Citerne Autre conditionnement : préciser

DECLARATION D'ELIMINATION DE DECHETS INDUSTRIELS

ENTREPRISE D'ELIMINATION		PERIODE
Raison sociale :　　　　　N°. SIRET :		TRIMESTRE :
Adresse :　　TRANSIT :　　ELIMINATION :		ANNEE :
Tél. :　　Nom du responsable :　　Visa :		

PRODUCTEUR (1) DU DECHET	DESIGNATION DU DECHET	CODE NOMENCLA-TURE R | A	Qté en tonnes	COLLECTEUR (1)	MODE DE TRAITEMENT	DESTINATION ULTERIEUR. DU DECHET (1) (3)

(1) raison sociale et localisation

(2) en cas de simple transit, préciser s'il y a reconditionnement du déchet et sous quelle forme

(3) ou le cas échéant, un pays étranger

Proposed trip-ticket

BORDEREAU DE SUIVI DE DECHETS INDUSTRIELS

A PRODUCTEUR

RAISON SOCIALE :

Adresse :

Téléphone :

Télex :

Responsable :

Atteste l'exactitude des renseignements ci-dessous, que les matières sont admises au transport selon les dispositions du règlement du 15.04.1945 et que notamment les conditions exigées pour le conditionnement et l'emballage ont été remplies.

Date de remise au transport :

VISA :

Quantité remise au transport
T

DESIGNATION DU DECHET	Code nomenclature(2) C | A	(1) Nom de la matière d'assimilation 1)	N° de groupe

CONSISTANCE DU DECHET
- ☐ Solide
- ☐ Blocs
- ☐ Granulés ou poudre
- ☐ Boue
- ☐ Pompable
- ☐ Pompable rechauffée
- ☐ Pelletable
- ☐ Liquide

TRANSPORT EN
- ☐ fûts nombre :
- ☐ benne
- ☐ citerne
- ☐ autre Précisez :
- ☐ bonbonne nombre :

ELIMINATION FINALE DU DECHET

Installation prévue :

Adresse :

N° du certificat d'acceptation préalable :

B COLLECTEUR - TRANSPORTEUR

RAISON SOCIALE :

Adresse :

Téléphone :

Ayant pris connaissance des indications ci-dessus,

Date :   VISA :

STOCKAGE

Oui Lieu:

Non

Quantité transportée :

T

C DESTINATAIRE

RAISON SOCIALE :

Adresse :

Téléphone :

Télex :

Responsable :

Refus de prise en charge le :

Motifs :

VISA :

Déchets pris en charge le :

En vue de l'opération désignée ci-dessous:

VISA :

Quantité reçue :
T

OPERATION PREVUE SUR LE DECHET
- ☐ Valorisation
- ☐ Détoxication
- ☐ Autre :
- ☐ Regroupement
- ☐ Incinération
- ☐ Mise en décharge
- ☐ Prétraitement

En cas de regroupement :

N° de cuve :

Destination finale du déchet :

En cas de prétraitement :

Description du prétraitement :

Destination finale du déchet :

(A remplir par le producteur)

En cas de transfert transfrontaliers :

Visa de la douane en cas d'exportation en dehors de la communauté et en cas de transit à travers la communauté.

Le déchet ci-dessus a quitté la communauté le:

Signature :   Cachet :

Etats de transit prévus, membres de la C.E.E :	Bureaux de douane d'entrée correspondants :
1	1
2	2
3	3
4	4
5	5
6	6

(1) Au titre du R.T.M.D

(2) Selon la nomenclature établie par le S.E.E.Q.V.

(3) Si le nombre de déchets mélangés est supérieur à 2, utiliser un (ou plusieurs) bordereau(x) supplémentaire(s)

FRANCE

Notification system

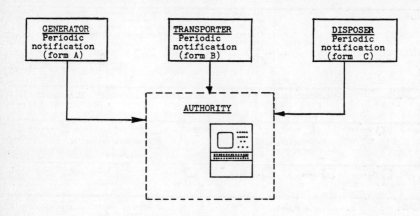

Trip ticket system

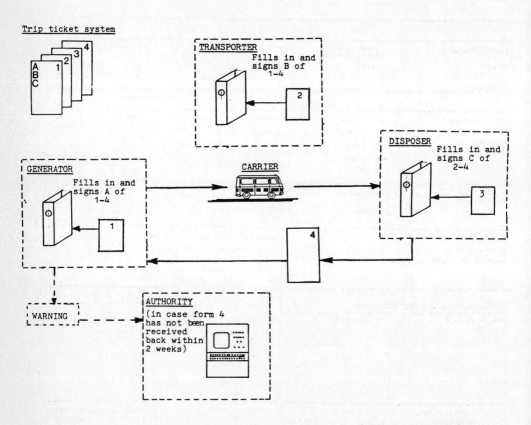

Annex 2

GERMANY

Notification and trip-ticket system

The notification and manifest system is a combined and integrated system.

The basis of the system is as follows:

Before transport can be accomplished there must be a valid contract between the three concerned parties: generator, transporter and disposer. The manifest must be completed before transport begins.

The trip-ticket (Begleitschein) has 6 copies.

The procedure is as follows:

- All copies filled in by the generator
- Copy 1-6 signed by the generator and the transporter
- Copy 1 is for the record-book of the generator
- Copy 2 must be sent by the generator to the authority
- Copy 3, 4, 5 and 6 accompany the transport
- Copy 3, 4, 5 and 6 must be signed by the disposer
- Copy 3 is for the record-book of the transporter
- Copy 4 is for the responsible authority
- Copy 5 must be sent back to the generator as evidence that the waste reached its destination
- Copy 6 is for the record-book of the disposer.

On the form, information must provided concerning the parties of interest (generator, transporter and disposer), about the waste (type, code-number, quantity in m^3 and in tonnes), means of transport and relevant data (appropriate declaration of load, date of transport and date of receipt of the waste). Each form has to be signed by all parties.

GERMANY

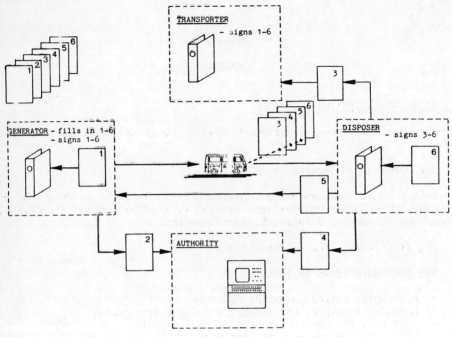

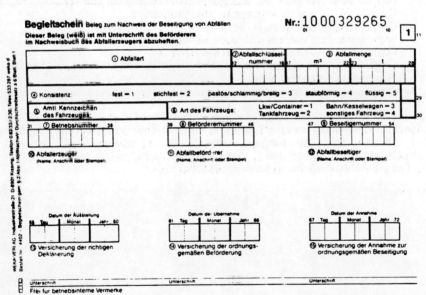

Annex 3

THE NETHERLANDS

Notification forms

Only a notification form is needed. For transport there is no special (waste) manifest system.

For notification, two procedures are possible:

- for the first time a waste is reported

- for the regular reporting of a waste which has already been reported to the government authorities.

The procedure for a new waste type or a new agreement between generator and disposer is as follows:

- the generator completes 4 copies of declaration form C (C_1-C_2-C_3-C_4) and signs them.

    (This declaration-form will be the basis for the disposal-agreement with the licensed disposer).

- the generator sends C_1-C_2-C_3 to the disposer; C_4 is retained for his files.

- the disposer accepts the note (or not), signs C_1-C_2-C_3 and gives the waste a "waste streamnumber".

- the disposer sends C_1 back to the generator, C_3 is for his files and C_2 remains for the time being at the disposer. (Now the generator knows that his waste can be treated).

- next the generator completes form A, together with the "waste streamnumber" given by the disposer and signs copies (A_1-A_2).

- within 8 days after transport, the generator sends A_1 to the government Department of Environment) while A_2 is for his files.

    (Now the government is notified that the waste will be shipped or has already been shipped).

- the disposer sends form C_2 to the government as evidence that the waste arrived at this plant and was accepted.

For regular transports of the same waste from the same generator to the same disposer, treated in the same way, the procedure is as follows:

- the generator completes B_1-B_2.

By use of the "waste streamnumber" and the usual name of the waste, the type of waste is identified as well as the disposer.

Quarterly the generator sends B_1 to the government and retains B_2 in his files.

- the disposer completes D_1-D_2.

The disposer sends D_1 monthly to the government and retains D_2 for his files.

In the case of declaration of a new waste, the required information is very detailed. However, when the the waste is shipped on a regular basis this detailed information is unnecessary.

The Netherlands

A new waste which is shipped for the first time.

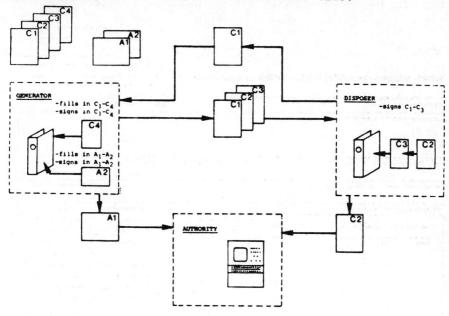

PROCEDURES FOR REGULAR WASTE STREAMS (monthly for B' and D')

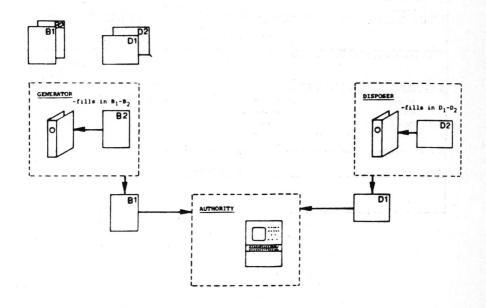

109

Form A

AFGIFTEMELDING CHEMISCHE AFVALSTOFFEN formulier A (wit)

insturen binnen acht dagen na de datum van de afgifte
aan de Minister van Volksgezondheid en Milieuhygiëne,
t.a.v. de sector Afvalstoffen
postbus 439
2260 AK LEIDSCHENDAM

vul voor chemische afvalstoffen die naar aard, eigenschappen of samenstelling verschillen,
afzonderlijke formulieren in.

2a. naam en adres van de aanbieder	b. bedrijfstak	telefoon	telex
		kontaktpersoon	

3a. gebruikelijke naam van het afval	b. datum van afgifte jaar maand dag	c. hoeveelheid per afgifte kg

4a. van (welke fase in) welk proces is het afval afkomstig?	b. procesnummer *

5a. samenstelling van het afval *

naam en formule van de component	stofnummer *	gewichtspercentage * gemiddeld	maximaal	b. aggregatiebetoestand **	c. zuurgraad **
water H₂O	0 0 0 1			1 poeder	vaste stof of poeder
				2 vaste stof	1 sterk zuur
				3 steekvast, pasta	2 zwak zuur
				4 slurry	3 neutraal
				5 dikke vloeistof	4 zwak basisch
				6 dunne vloeistof	5 sterk basisch
				7 (vloeibaar) gas	vloeistof pH
				d. aard **	e. volumieke massa kg/m³
				1 explosief	
				2 agressief	f. stookwaarde J/kg
				3 corrosief	
				4 giftig	g. kookpunt °C
				h. vlampunt °C	i. smeltpunt °C
		100		bepalingsmethode	j. zelfontbrandingstemp °C

6a. het afval is vervoerd door **	3 een derde nl. (naam en adres)	b. adres van aflevering van het afval
1 de aanbieder		
2 de verwerker		

c. soort en maat verpakking	d. vervoer over **	e. gevaar- en stofidentificatiecodes *
	1 weg 2 spoor 3 water	(Wet gevaarlijke stoffen)

7. naam en adres van degene aan wie het afval is afgegeven

8. kunt u aangeven op welke wijze het afval wordt verwijderd? **

1 destilleren	11 lozen op zee	21 anders nl.
2 extraheren	12 verbranden op zee	
3 ONO (= 4 + 5 + 6) 4 ontgiften	13 verbranden	
5 neutraliseren	14 op of in de bodem brengen	
6 ontwateren	15 bewaren	

plaats en datum   handtekening van de aanbieder

ieder kwartaal op apart formulier

insturen binnen acht dagen na afloop van het
betreffende kalenderkwartaal

aan: de Minister van Volksgezondheid en Milieuhygiëne,
t.a.v. Directie Afvalstoffen en Schone Technologie,
postbus 439
2260 AK LEIDSCHENDAM

PERIODIEKE AFGIFTEMELDING CHEMISCHE AFVALSTOFFEN
formulier B (blauw) over   kwartaal

naam en adres van de aanbieder					telefoon	telex
					kontaktpersoon	

datum van afgifte			afvalstroomnummer*		hoeveelheid	
jaar	maand	dag	code	verwerkingsnummer	in kilogram	gebruikelijke naam van het afval
1						
2						
3						
4						
5						
6						
7						
8						
9						
10						
11						
12						
13						
14						
15						
16						
17						
18						
19						
20						
21						
22						
23						
24						
25						
26						
27						
28						
29						

	afvalstroomnummer		mutatie(s)**
	code	verwerkingsnummer	
30			
31			
32			
33			
34			
35			

plaats en datum	handtekening

Form C

OMSCHRIJVING CHEMISCHE AFVALSTOFFEN formulier C (groen)

I nsturen zonder de laatste kopie naar de verwerker

II na acceptabe dit groene formulier naar de aanbieder retour

vul voor chemische afvalstoffen die naar aard, eigenschappen of samenstelling erschillen, afzonderlijke formulieren in.

2a. naam en adres van de aanbieder	b. bedrijfstak	telefoon	telex
		kontaktpersoon	

3a gebruikelijke naam van het afval	b hoe vaak zal afgifte plaatsvinden	c. hoeveelheid per afgifte kg

4a. van (welke fase in) welk proces is het afval afkomstig?	b. procesnummer *

5a samenstelling van het afval *					b. aggregatietoestand **	c. zuurgraad **

naam en formule van de component	stofnummer *	gewichtspercentage *				

gewichtspercentage: gemiddeld / maximaal

aggregatietoestand:
1 poeder
2 vaste stof
3 steekvast, pasta
4 slurry
5 dikke vloeistof
6 dunne vloeistof
7 (vloeibaar) gas

zuurgraad: vaste stof of poeder
1 sterk zuur
2 zwak zuur
3 neutraal
4 zwak basisch
5 sterk basisch
vloeistof pH:

naam en formule van de component	stofnummer *	gemiddeld	maximaal
water H2O	0 0 0 1		
	100		

d. aard**
1 explosief
2 agresief
3 corrosief
4 giftig

h. vlampunt °C
bepalingsmethode

e. volumieke massa kg/m³
f. stookwaarde J/kg
g. kookpunt °C
i. smeltpunt °C
j. zelfontbrandingstemp °C

6a. het afval wordt vervoerd door **	3 een derde nl. (naam en adres)	b. adres van aflevering van het afval
1 de aanbieder		
2 de verwerker		

c. soort en maat verpakking	d. vervoer over ** 1 weg 2 spoor 3 water	e. gevaar- en stofidentificatiecodes * (Wet gevaarlijke stoffen)

de aanbieder verklaart zich akkoord met de verwerkingsvoorwaarden

plaats en datum   handtekening van de aanbieder

acceptabe door de verwerker

plaats en datum   handtekening van de verwerke

112

Form D

insturen binnen acht dagen na afloop van de
betreffende kalendermaand

ONTVANGSTMELDING CHEMISCHE AFVALSTOFFEN
formulier D (rood)

aan: de Minister van Volksgezondheid en Milieuhygiëne.
t.a.v. de sector Afvalstoffen
postbus 439
2260 AK LEIDSCHENDAM

naam en adres van de verwerker					telefoon		telex
					kontaktpersoon		

datum van ontvangst			afvalstroomnummer*		hoeveelheid in kilogram	gebruikelijke naam van het afval	°°° code
jaar	maand	dag	code	verwerkingsnummer			

	afvalstroomnummer		mutatie(s)°°	
	code	verwerkingsnummer		

plaats en datum .... handtekening ....

Annex 4

SWEDEN

No manifest system is yet in practice. The procedure in mind is as follows:

-- The forms for the hazardous waste are filled in. A copy is reserved for the community environment- and health authority (4) and one is kept with the producer (5). Copies (1), (2) and (3) are sent with the transport and (1) is kept with the treatment company. Number (3) goes for the file of the transporting company.

-- A preliminary data processing or manual listing takes place at the treatment company. Information is transmitted to a central computer either directly via the telephone system or is transferred to data disquettes which are sent by mail.

-- Data coming from different treatment companies are processed. Data lists are printed for the environment- and health authorities and for the county administrations.

-- The content of the compiled lists of data are compared with a selection of copies (4) from the waste producers.

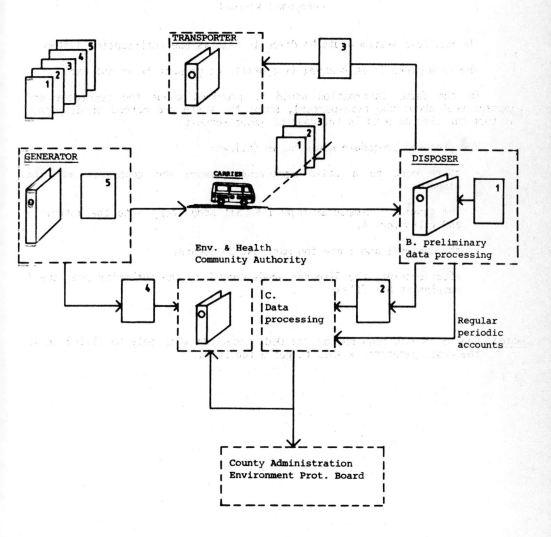

Annex 5

SWITZERLAND
(proposed system)

The manifest system would be directly used as the notification system.

The form given in this annex is a draft. At present it is not used.

On the form, information would be provided about the generator and disposer (not about the transporter), about the waste and method of disposal. The form can also be used in the case of import-export.

The proposed procedure would be as follows:

- there must be a valid agreement between the generator and the disposer

- the generator completes copy 1-4 (5); sends copy 3 to the authority and files copy 4.

- the copies 1 and 2 are the manifest documents.

- after receipt, the disposer sends copy 2 to the authority and copy 1 remains in his files.

Note: Copy 5 is not required by the Ordinance. It will only be filled in if the waste generator wishes to get a feedback.

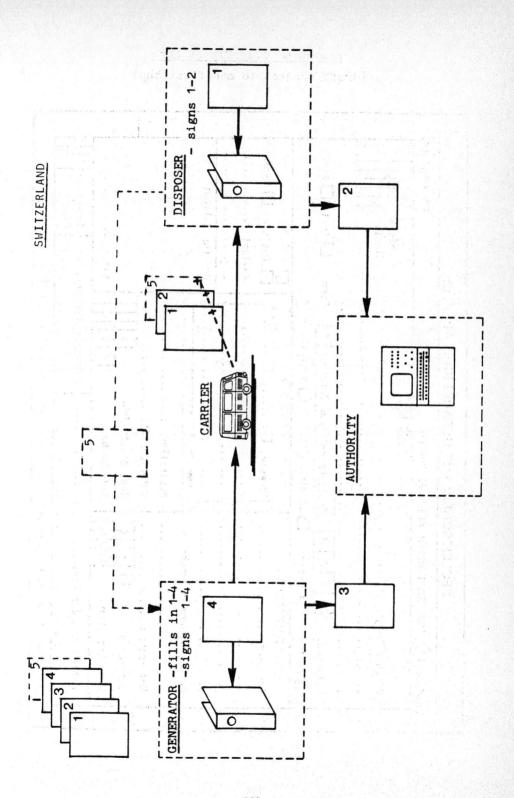

SWITZERLAND

DISPOSER — signs 1-2

CARRIER

AUTHORITY

GENERATOR — fills in 1-4
— signs 1-4

Sketch of Trip-Ticket Form

(draft subject to modifications)

BEGLEITSCHEIN FÜR GEFÄHRLICHE ABFÄLLE

LAUF-NR.06

KOPIE ① VOM EMPFÄNGER AUFZUBEWAHREN

ABFALL

Umschreibung: ....

Gefahren: ⬚ P ⬚ A
⬚ pastos / ⬚ flüssig/ ⬚ Ausserordentliche Gefahren !
⬚ in Bindemittel / ⬚ verfestigt

Form: ⬚ fest/ ⬚ Staub

Code-Nr.: ....
Menge: ....   kg
                            l

EMPFÄNGER

Adresse: ....

Entgegengenommen durch: ....

Unterschr.: ....
Betrieb Nr. ....
Beseitigungsart   1 2 3 4 5 6

ABGEBER

Adresse: ....

Aussteller: ....
Unterschrift: ....
Tag, Monat, Jahr
Betrieb Nr.

⬚ Inland
⬚ Ausfuhr ⬚ Einfuhr
Zollamt: ....

Res. für Zollamt

Vis: ....

Bezeichnung der Gebinde: ....

Antrag/Bewilligung Nr:

Transport
Code

Eintragungen bitte mit Stempel, Schreibmaschine oder in Blockschrift

118

Annex 6

UNITED KINGDOM

Notification and manifest are combined into one system. The WDA must be notified before transport occurs.

On the document, information is given about the generator, the transporter and the disposer, about the waste, the date of transport and delivery to the waste disposer. No information is given about the method of disposal.

The procedure is as follows:

- the generator completes parts A and B copies 1-6 (including who the disposer is) and signs part A

- the generator sends copy 1 to the Waste Disposal Authority in whose area the waste will be disposed (at least 3 days in advance)

- on collection, the transporter completes and signs part C of copies -5; the producer fills in and signs part D of copies 2-5

- the generator sends copy 2 to his own WDA and retains copy 3

- the copies 4-6 are the trip-ticket documents

- the disposer completes and signs part E of copies 4-6 on receipt of the waste;
  the disposer sends copy 4 to the transporter, copy 5 to his own WDA and retains copy 6.

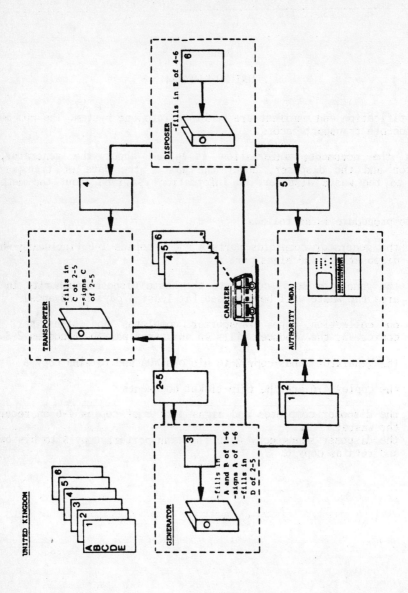

Department of the Environment/Welsh Office/Scottish Development Department

CONSIGNMENT NOTE FOR THE CARRIAGE & DISPOSAL OF HAZARDOUS WASTES

Serial No

PRENOTIFICATION COPY

Producer's Certificate **A**	(1) The material described in B is to be collected from and (2) taken to Signed Name On behalf of Position Address and telephone Date Estimated date of collection
Description of the Waste **B**	(1) General description and physical nature of waste (2) Relevant chemical and biological components and maximum concentrations (3) Quantity of waste and size, type and number of containers (4) Process(es) from which waste originated
Carrier's Collection Certificate **C**	I certify that I collected the consignment of waste and that the information given in A(1) & (2) and B(1) & (3) is correct, subject to any amendment listed in this space: I collected this consignment on at hours Signed Name Vehicle Registration No On behalf of Address and telephone Date
Producer's Collection Certificate **D**	I certify that the information given in B & C is correct and that the carrier was advised of appropriate precautionary measures. Signed Name Telephone Date
Disposer's Certificate **E**	I certify that Waste Disposal Licence No , issued by County/District Council, authorises the treatment/disposal at this facility of the waste described in B (and as amended where necessary at C). Name and address of facility This waste was delivered in vehicle (Reg No) at hours on (date) and the carrier gave his name as on behalf of Proper instructions were given that the waste should be taken to Signed Name Position Date on behalf of
For use by Producer/ Carrier/ Disposer	

Annex 7

UNITED STATES

The procedure is as follows:

- the generator makes a contract with the disposing firm and identifies the transporter

- the generator completes the manifest and retains a copy for his own files

- the other copies go to the transporter (1) and the designated disposer (2)

- the transporter and disposer retain a copy for their files

- the disposer returns, after accepting the waste, a copy to the generator

(No copy is sent to EPA)

In the USA a biannual report is required, but not on a shipment by shipment basis.

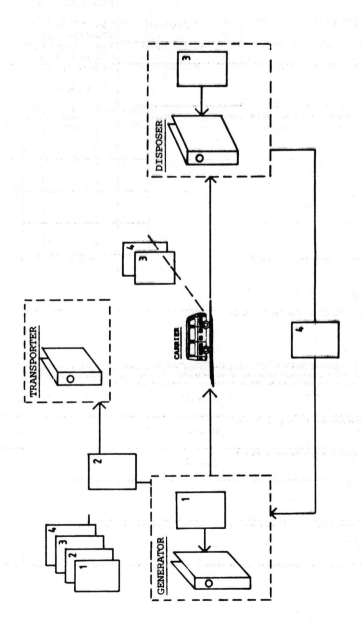

UNIFORM HAZARDOUS WASTE MANIFEST	1. Generator's US EPA ID No. Manifest Document No.	2. Page 1 of	Information in the shaded areas is not required by Federal law.

3. Generator's Name and Mailing Address	A. State Manifest Document Number
	B. State Generator's ID
4. Generator's Phone ( )	

5. Transporter 1 Company Name	6.	US EPA ID Number	C. State Transporter's ID
			D. Transporter's Phone
7. Transporter 2 Company Name	8.	US EPA ID Number	E. State Transporter's ID
			F. Transporter's Phone
9. Designated Facility Name and Site Address	10.	US EPA ID Number	G. State Facility's ID
			H. Facility's Phone

G E N E R A T O R	11. US DOT Description (Including Proper Shipping Name, Hazard Class, and ID Number)	12. Containers		13. Total Quantity	14. Unit Wt/Vol	I. Waste No.
		No.	Type			
	a.					
	b.					
	c.					
	d.					

J. Additional Descriptions for Materials Listed Above	K. Handling Codes for Wastes Listed Above

15. Special Handling Instructions and Additional Information

16. **GENERATOR'S CERTIFICATION:** I hereby declare that the contents of this consignment are fully and accurately described above by proper shipping name and are classified, packed, marked, and labeled, and are in all respects in proper condition for transport by highway according to applicable international and national governmental regulations.

Printed/Typed Name	Signature	Date
		Month Day Year

T R A N S P O R T E R	17. Transporter 1 Acknowledgement of Receipt of Materials		Date
	Printed/Typed Name	Signature	Month Day Year
	18. Transporter 2 Acknowledgement or Receipt of Materials		Date
	Printed/Typed Name	Signature	Month Day Year

F A C I L I T Y	19. Discrepancy Indication Space
	20. Facility Owner or Operator: Certification of receipt of hazardous materials covered by this manifest except as noted in Item 19.

Printed/Typed Name	Signature	Date
		Month Day Year

EPA Form 8700-22 (3-84)

UNIFORM HAZARDOUS WASTE MANIFEST *(Continuation Sheet)*	21. Generator's US EPA ID No.	Manifest Document No.	22. Page	Information in the shaded areas is not required by Federal law.

23. Generator's Name	L. State Manifest Document Number
	M. State Generator's ID

24. Transporter ____ Company Name	25. US EPA ID Number	N. State Transporter's ID
		O. Transporter's Phone
26. Transporter ____ Company Name	27. US EPA ID Number	P. State Transporter's ID
		Q. Transporter's Phone

28. US DOT Description *(Including Proper Shipping Name, Hazard Class, and ID Number)*	29. Containers No.	Type	30. Total Quantity	31. Unit Wt/Vol	R. Waste No.
a.					
b.					
c.					
d.					
e.					
f.					
g.					
h.					
i.					

S. Additional Descriptions for Materials Listed Above	T. Handling Codes for Wastes Listed Above

32. Special Handling Instructions and Additional Information

33. Transporter ____ Acknowledgement of Receipt of Materials		Date
Printed/Typed Name	Signature	Month Day Year

34. Transporter ____ Acknowledgement of Receipt of Materials		Date
Printed/Typed Name	Signature	Month Day Year

35. Discrepancy Indication Space

EPA Form 8700-22A (3-84)

Annex 8

CANADA

Below a short description is given of the coming Transportation of Dangerous Goods Act of 1980 and Regulations of 1984 in relation to the management of hazardous wastes.

Those individuals and circumstances that are exempt from the general provisions of the Regulations are set out in Part II of the Regulations (a manufacturer or user of the goods who is using them for the purposes of manufacture or use on property owned or leased by himself is, for example, exempt). It should however be noted that the Act and its Regulations will not apply to prevent transportation when the goods must be moved immediately for the safety of the public or the protection of the environment.

"Prescribed dangerous goods" are deemed (in Part III) to be those for which the shipment name, including any descriptive wording, the product identification number, the primary class, the division, if any, the subsidiary class, or classes, if any, or packing or compatibility group, if any, are listed. "Non-prescribed dangerous goods" are those products, substances or organisms that correspond to a listed shipping name but for which the class, division, packing group or compatibility group, or each or any one of these, must be ascertained. Nine classes of dangerous goods have been delineated and appear as follows:

(i)   Class 1 - explosives
(ii)   Class 2 - gases
(iii) Class 3 - flammable liquids
(iv)   Class 4 - flammable solids, spontaneously combustible materials and dangerous-when-wet materials
(v)   Class 5 - oxidizing materials and organic peroxides
(vi)   Class 6 - dangerous goods: poisonous and infectious substances
(vii) Class 7 - radioactive materials
(viii) Class 8 - corrosives
(ix)   Class 9.1 - miscellaneous substances
            9.2 - environmentally hazardous substances
            9.3 - waste streams

The documentation requirements as set out in Part IV of the Regulations include a number of exemptions (with regard to limited quantities, for example). But in general every consignor of dangerous goods shall (before transport) complete and sign a declaration. Additional documentation may be required if the consignment is being transported from a place in Canada to a place outside Canada other than the United States. Each declaration shall contain: (i) a description of the dangerous goods; (ii) the names and addresses of the consignee and consignor (and the person responsible for the loading facility when not the consignor); (iii) any special instructions to

A - CONSIGNOR (GENERATOR) EXPÉDITEUR

Provincial No. - N° provincial		

Company Name - Nom de l'entreprise

Circulation No. - N° de circulation

Address - Adresse

Shipping Site Address - Origine de l'expédition

City - Ville	Prov.	Postal Code - Code postal

Intended Consignee / Destinataire choisi

Provincial No. - N° provincial

Address - Adresse

Receiving Site Address - Destination de l'expédition

City - Ville	Prov.	Postal Code - Code postal

B - CARRIER TRANSPORTEUR

Company Name - Nom de l'entreprise — Provinc[ial]

Address - Adresse

City - Ville	Prov.

Registration / Immatriculation des véhicules	License No. / Immatriculation
Vehicle - Véhicule moteur	
Trailer #1 - Remorque #1	
Trailer #2 - Remorque #2	

Point of Entry / Point d'entrée	Point of Exit / Point de sortie

Carriers Certification: I declare that I have received the wastes described in Part the Intended Consignee

Déclaration du transporteur: Je declare avoir pris livraison des déchets décrits de les transporter au destinataire choisi

Name of Driver (Print) / Nom du conducteur (caractères d'imprimerie)	Tel. No. (Area C[ode]) / N° de tél. (ind. #)

Signature

Physical State (Sol., Liq., Gas) État physique (sol., liq., gaz)	Shipping Name Appellation réglementaire	Product Identification No. N° d'identification du produit		Classification	Packing Group / Groupe d'emballage	Quantity Shipped Quantité expédiée		Concentration	
		Provincial	TDGA LTMD				Units L/kg Unités		Units (Specify) Unités (spécifier)
		⌐⌐⌐⌐⌐	⌐⌐⌐⌐⌐						
		⌐⌐⌐⌐⌐	⌐⌐⌐⌐⌐						
		⌐⌐⌐⌐⌐	⌐⌐⌐⌐⌐						
		⌐⌐⌐⌐⌐	⌐⌐⌐⌐⌐						
		⌐⌐⌐⌐⌐	⌐⌐⌐⌐⌐						

Special Handling/Emergency Instructions / Manutention spéciale/Instructions d'urgence   Attached / Ci-jointes ☐   Below / Plus bas ☐

Shipped - Date d'expédition / Time - Heure   Date (D/M/Y - J/M/A)   AM ☐ PM ☐

Scheduled Arrival Date (D/M/Y) / Date prévue d'arrivée (J/M/A)

Consignor Certification: I declare that the information contained in Part A is correct and complete.

Déclaration de l'expéditeur: Je declare que tous les renseignements à la partie A sont véridiques et complets.

Name of Authorized Person (Print) / Nom de l'agent autorisé (caractères d'imprimerie)	Tel. No. (Area Code) / N° de tél. (ind. rég.)	Signature

(1-84)

MANIFEST - MANIFESTE

Manifest Reference No.
N° de référence du manifeste **A000007**

N° provincial

Reference No.'s of Other Manifest(s) used
N°'s de référence des autres manifestes utilisés

Code - Code postal

Province

C - CONSIGNEE (RECEIVER)
DESTINATAIRE

EMERGENCY TELEPHONE
NUMBERS
N°S DE TÉLÉPHONE EN CAS
D'URGENCE

Company Name - Nom de l'entreprise | Provincial No. - N° provincial

Canutec (Call Collect)
Canutec (appeler à frais virés)

Address - Adresse

(613)-966-6666

Alberta

City - Ville | Prov. | Postal Code - Code postal

1-800-222-6514

very to
e A afin

Receiving Site Address - Destination de l'expédition

British Columbia
Colombie-Britannique

City - Ville | Prov. | Postal Code - Code postal

(604) 387-5956

Manitoba

Received - Réception
Time - Heure   Date (D/M/Y J/M/A)

AM ☐   PM ☐

(204) 944-4888

New Brunswick
Nouveau-Brunswick

kaging nants	Code (Int/Ext)	Quantity Received Quantité reçue	Units L/kg Unités	Identify any Shipment Discrepancy/Problems/ Refusal Identifier toute différence entre manifeste et cargaison/ problemes/refus	Handling Code Code de manutention	Decontamination - Décontamination			
						Packaging Contenants		Vehicle Véhicule	
						Yes Oui	No Non	Yes Oui	No Non

Zenith 4-9000

Newfoundland
Terre-Neuve

(709)-772-2083

Northwest Territories
Territoires du Nord-Ouest

(403)-920-8130

Nova Scotia
Nouvelle-Écosse

Zenith 4-9000

Ontario

If Handling Code "Other" (Specify)
Si code de manutention "divers", spécifier

If waste to be re-transferred (specify company name)
Si le déchet doit être re-transféré, indiquer le nom de l'entreprise | Provincial No. N° provincial

Prince Edward Island
Île du Prince-Édouard

Zenith 4-9000

Quebec
Québec

Address - Adresse | City - Ville | Prov.

(418)-643-4595

Saskatchewan

Consignee Certification: I declare that the information contained in Part C is correct and complete
Déclaration du destinataire: Je déclare que tous les renseignements à la partie C sont véridiques et complets

Name of Authorized Person (Print)
Nom de l'agent autorisé (caractères d'imprimerie)

1-800-667-3503

Yukon Territory
Territoire du Yukon

Tel. No. (Area Code) - N° de tél. (ind. rég.) | Signature

(403)-667-7244

ace of Consignee - (mailed by Consignor) - Province du destinataire - (Postée par l'expéditeur)   **1**

ensure safe handling and storage during transportation; (iv) the total mass or volume and (v) the words "SPECIAL DANGEROUS GOODS" when required. In the case of hazardous wastes the declaration shall be termed a "waste manifest" and will include such information as the destination of the wastes where it is not the same as the consignee and the name and telephone number of each carrier that is expected to transport the wastes, for example. The consignor, carrier or carriers and consignee of dangerous goods shall each retain a copy of the declaration in Canada for at least two years after the goods have been delivered and accepted by the consigner.

Where a consignment of hazardous wastes is to be transported, the consignor shall send one copy of the manifest relating to the wastes to the appropriate authority of the places or origin and destination within two days after delivery of wastes to the carrier. Every carrier shall ensure that the declaration is marked with the date of delivery to him. When a consignment of hazardous wastes is received by the consignee that person shall: (i) on receipt of the hazardous wastes, complete and sign and date the copies of the manifest that accompany the wastes indicating thereon whether or not the included information accurately reflects the wastes in the consignment at the time the consigment reached its destination; (ii) send one copy of the manifest relating to the wastes to each of the appropriate authorities of the places of origin and destination within two working days after the wastes have reached their destination; (iii) retain, in Canada, one copy of the manifest for at least two years after the wastes have reached their destination; (iv) send one signed copy of the manifest to the carrier within two days after the wastes have reached their destination. The carrier of a consignment of hazardous wastes shall retain one copy of the manifest in Canada for at least two years after he has delivered the wastes to the consignee.

Part V of the Regulations deals with safety markings (labels, placarding requirements, colour standards and identification marks are among the topics reviewed). Safety requirements (dealing with such matters as training, for example) are similarly discussed in later sections.

Annex 9

AUSTRIA

The notification and manifest system is a combined and integrated system.

The trip-ticket (Begleitschein) has 6 copies.

The procedure is as follows:

-- All copies filled in by the generator
-- Copy 1-6 signed by the generator
-- Copy 1 is for the record-book of the generator
-- Copy 2 is as evidence that the waste is collected by the transporter
-- Copy 3 must be sent by the transporter to the authority
-- Copies 4, 5, 6 accompany the transport
-- Copies 4, 5, 6 must be signed by the disposer
-- Copy 4 is for the record-book of the transporter
-- Copy 5 is for the record-book of the disposer
-- Copy 6 is for the responsible authority.

On the form, information must be provided concerning the parties of interest (generator, transporter and disposer), about the waste (type, code-number, quantity in m^3 and in tonnes), means of transport and relevant data (appropriate declaration of load, date of transport and date of receipt of the waste).

The generator signs copies 1 - 6
The transporter signs copies 2 - 6
The disposer signs copies 4 - 6

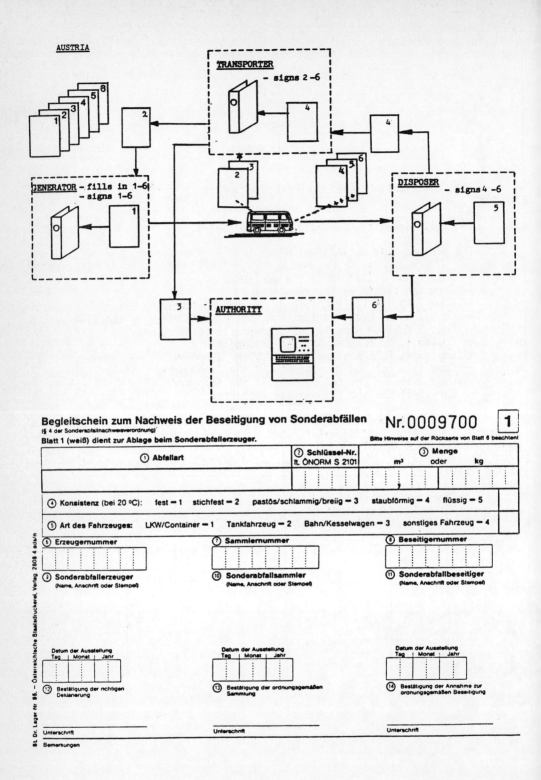

AUSTRIA

TRANSPORTER — signs 2 –6

GENERATOR – fills in 1–6 / – signs 1–6

DISPOSER — signs 4 –6

AUTHORITY

Begleitschein zum Nachweis der Beseitigung von Sonderabfällen   Nr. 0009700 **1**

(§ 4 der Sonderabfallnachweisverordnung)

Blatt 1 (weiß) dient zur Ablage beim Sonderabfallerzeuger.   Bitte Hinweise auf der Rückseite von Blatt 6 beachten!

① Abfallart	② Schlüssel-Nr. lt. ÖNORM S 2101	③ Menge		
		m³	oder	kg

④ Konsistenz (bei 20 °C):   fest — 1   stichfest — 2   pastös/schlammig/breiig — 3   staubförmig — 4   flüssig — 5

⑤ Art des Fahrzeuges: LKW/Container — 1   Tankfahrzeug — 2   Bahn/Kesselwagen — 3   sonstiges Fahrzeug — 4

⑥ Erzeugernummer   ⑦ Sammlernummer   ⑧ Beseitigernummer

⑨ Sonderabfallerzeuger (Name, Anschrift oder Stempel)   ⑩ Sonderabfallsammler (Name, Anschrift oder Stempel)   ⑪ Sonderabfallbeseitiger (Name, Anschrift oder Stempel)

Datum der Ausstellung   Tag | Monat | Jahr

⑫ Bestätigung der richtigen Deklarierung   ⑬ Bestätigung der ordnungsgemäßen Sammlung   ⑭ Bestätigung der Annahme zur ordnungsgemäßen Beseitigung

Unterschrift   Unterschrift   Unterschrift

Bemerkungen

BL Dr. Lager Nr 95. – Österreichische Staatsdruckerei, Verlag 2808 4 aus/n

132

Annex 10

NORWAY

A system of declaration forms acting as trip tickets have been in operation for some years. The form is now revised into a combined trip ticket and manifest system.

The form is divided into three sections; one section is to be filled in by the waste generator, one section by the waste receiver or collector and one by the final disposer of the waste.

The waste generator fills in the first section with information on the waste, i.e. type, amount, origin, properties, composition of the waste.

The waste receiver is the enterprise which receives the waste directly from the waste generator. The receiver may be reception station or a treatment plant. On transportation to the waste receiver, all four copies of the form shall accompany the lot. The waste receiver gives a confirmation of the reception by filling out the second section of the form. Copy (4) is sent to the waste generator and copy (3) is filed by the receiver.

By transportation from the waste receiver to the final disposal plant copy (1) and (2) accompany the lot. The waste disposer fills in the third section of the form with information on the disposal. Copy (2) is kept by the disposal plant and copy (1) is sent to the environment authorities.

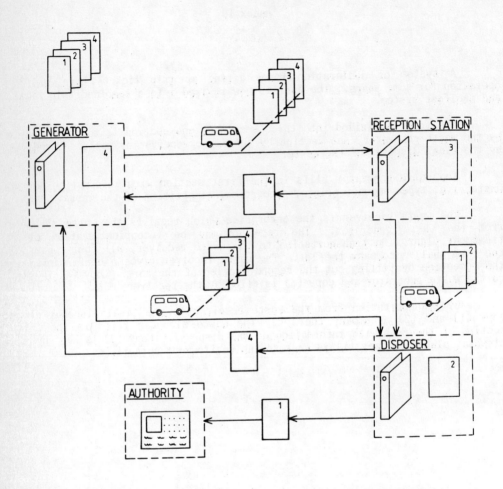

134

STATENS FORURENSNINGSTILSYN

DEKLARASJON AV SPESIALAVFALL

Fylles ut ved levering til mottaks- eller behandlingsanlegg for spesialavfall — En blankett for hver av felles type
Eksplosive, selvantennende, smittefarlige og radioaktive stoffer samt organiske peroksyder mottas ikke!
Ang. utfylling, leveringsbetingelser mv. — se veiledningen på baksiden

Deklarasjons nr.

1. Leverandør (Bedrift)	2. Gate/vei adresse (Bedriftens beliggenhet)

3.1 Postadresse	3.2 Postnr.	3.3 Poststed

4. Kontaktperson hos leverandør	Telefon	5. Bedriftens momsnr.	6. Kommune hvor avfallet har oppstått	6.2 Kommunenr.

7. Mengde avfall. 7.1 Antall kg	kg	7.2 Antall kolli	kolli	8. Tåler frost 1 ☐ Ja 2 ☐ Nei	9. Fysisk tilstand ved romtemperatur 1. ☐ Pumpbart 2. ☐ Fast

10. Fareklasse

2. ☐ Gass
3 ☐ Brannfarlig væske
4.1 ☐ Brannfarlig fast stoff

4.3 ☐ Stoff som gir brannfarlig gass ved kontakt med vann
5.1 ☐ Oksyderende stoff

6.1 ☐ Giftig stoff
8 ☐ Etsende stoff

11. Avfallsgruppe

1 ☐ Spillolje	10 ☐ Cyanidholdig stoff
2 ☐ Oljeavfall fra renseanlegg for oljeholdig avløpsvann	11 ☐ Kasserte plantevernmidler
3 ☐ Olje-emulsjoner	
4.1 ☐ Organiske løsningsmidler med halogen	Andre avfallsgrupper
4.2 ☐ Organiske løsningsmidler uten halogen	12 ☐ PCB-holdig avfall
5 ☐ Maling, lim, lakk og trykkfargeavfall	13 ☐ Isocyanater
6 ☐ Destillasjonsrester	14 ☐ Annet organisk avfall
7 ☐ Tjæreavfall	15 ☐ Sterke syrer
8 ☐ Kvikksølv- eller kadmiumholdig avfall	16 ☐ Sterke baser
9 ☐ Avfall som inneholder bly, kopper, sink, krom, nikkel, arsen, selen eller barium	17 ☐ Annet uorganisk avfall

12. Innhold av spesielle komponenter

	Ja	Nei	Prosent		Ja	Nei	Prosent
1 Kreft- eller allergifremkallende stoff	☐	☐		4 Tungmetaller	☐	☐	
2 Vann	☐	☐		5 Aske/sediment	☐	☐	
3 Klor, brom, fluor, jod	☐	☐					

13. Nærmere beskrivelse av avfallet: (Vedlegg gjerne eventuelle produktdatablad e.l.) Oppgi så mye som mulig om innhold av kjemiske komponenter og deres konsentrasjoner, kjemisk formel avfallets opprinnelse, hva det er forurenset med, navn og adresse til importør eller produsent osv.:

1 Handelsnavn

2 Prosess

3 Kjemisk sammensetning

4 Andre opplysninger

14 Dato	Leverandørens underskrift
Dag Mnd År	

Fylles ut av mottaker

15.1 Mottaker (navn/stempel)	15.2 Mottakers Reg.nr.	15.3 Mottaks-gebyr	Dato	Underskrift

Fylles ut av behandlingsanlegget

16 Behandlingsmate (endelig disponering)

1. ☐ Forbrenning med varmegjenvinning
2. ☐ Forbrenning uten varmegjenvinning
3. ☐ Materialgjenvinning
4. ☐ Deponering
5. ☐ Kjemisk behandling/utslipp

6 ☐ Eksport, mottaksland ....
7 ☐ Annet, spesifiser ....

17 Behandlings-anl. reg.nr.

Dato	Underskrift

135

Annex 11

EUROPEAN COMMUNITIES

The Council Directive on the Supervision and Control within the European Community of the Transfrontier Shipment of Hazardous Wastes within the European Community contains a notification document and a consignment document (manifest) for governing transfrontier shipments.

The proposed procedure is as follows:

-- the generator, who wishes to export, completes the notification document copies 1-4 and sends:

Copy 1 to the competent authority of the importing country

Copy 2 to the authority of the exporting country

Copy 4 to the proposed disposer

Copy 3 remains with the generator

-- the competent authority of the importing country sends a "receipt of acknowledgement of the notification" back (within one month)

-- (now the way is open for the prospective transboundary-transport)

-- the generator fills in the trip-ticket (8 copies): (additional information is given by the transporter)

Copy 1 for his own files

Copy 2 is to be sent to the authority of the exporting country

Copy 3 is to be sent to the authority of the importions country (in the case of transit a copy of copy 2 is to be sent to the transit country)

-- the remaining copies 4-8 are handed over to the transporter, together with a copy of the acknowledgement

-- the transporter hands over copies 5-8 to the disposer; copy 4 remains for his records

- the disposer sends copy 5 to the generator, copy 6 to the authority of the exporting country and copy 7 to the authority of the importing country, not later than 2 weeks after receipt of the waste (in the case of transit, a copy of copy 7 is to be sent to the transit country)

the disposer retains copy 8 for his records.

General notification may be given for a specified period if there are continuous shipments.

CONTENT OF THE UNIFORM CONSIGNMENT NOTE

SECTION A

Information to be provided on notification

1. Holder of the waste(a)
2. General notification or notification of a single shipment
3. (a) Consignee of the waste(a)
   (b) Permit No (where applicable)
   (c) Information relating to the contractual agreement between the holder and the consignee
4. Producer(c) of the waste(a)
5. (a) Carriers(s) transporting the waste(a)
   (b) Licence No (where applicable)(b)
6. (a) Country of origin of the waste
   (b) Competent authority(c)
7. (a) Expected countries of transit
   (b) Competent authority(c)
8. (a) Country of destination of the shipment
   (b) Competent authority(c)
9. Planned date(s) of shipment(s)(d)
10. Means of transport envisaged (road, air, sea, etc.)
11. Information relating to insurance against damage to third parties(f)
12. Name and physical description of the waste and its composition(c)
13. Method of packing envisaged
14. Quantit(y)(ies)(kg)(g)
15. UN classification
16. Process by which the waste was generated
17. Nature of the risk: Explosive/Reactive/Corrosive/Toxic/Flammable/Other
18. Outward appearance of the waste at ...°C; Powdery or Pulverulent/Solid/Viscous or Syrupy/Sludgy/Liquid/Gaseous/Other
    Colour
19. Place of generation of the waste
20. Place of disposal of the waste
21. Method of disposal of the waste
22. Other information
23. Declaration by the holder that the information is correct, place, date, signature of holder.

(a) Full name and address, telephone and telex number and the name, address, telephone or telex number of the person to be contacted.
(b) Where there is no specific licence, the carrier should be able to demonstrate that the complies with the rules of the Member State concerned in respect of transport of such waste.
(c) Full name and address, telephone and telex number. This information is obligatory only where the countries concerned are Member States. In other cases it should be provided if known.
(d) In the case of a general notification covering several shipments, either the expected dates of each shipment or, if this is not known, the expected frequency of the shipments will be required.
(e) The nature and the concentration of the most characteristic components, in terms of the toxicity and other dangers presented by the waste will be required together with, if possible, an analysis referring to the method of disposal envisaged, particularly in the case of an initial shipment.
(f) Examples of information to be included where such insurance is required: insurer, policy number, last day of validity.
(g) In the case of a general notification covering several shipments, both the total quantity and the quantities for each individual shipment will be required.

SECTION B
Acknowledgement
(see General Instructions - paragraph 2)

1. Date of receipt of notification
2. Date on which acknowledgment is sent
3. Period of validity of acknowledgement
4. Whether acknowledgement applies to a single shipment or to several shipments
5. Date, signature and stamp of competent authority.

SECTION C
Transport arrangements
(See General Instructions - paragraph 3)

1. Serial N° of shipment
2. Identification of means of transport
3. N° and type of containers, markings, numbers, etc.
4. Exact quantities (kg)
5. Customs posts of entry in the country(ies), whose territory is to be passed through
6. Special conditions (if any) set by Member States concerned regarding the transport across their territory
7. Declaration by the holder and the carrier that the information is correct; place, date, signature of holder and carrier.

SECTION D
Receipt by the consignee
(See General Instructions - paragraph 4)

Declaration by the consignee that he has received the waste for disposal and the quantity thereof; place, date, signature of the consignee.

SECTION E
Customs endorsement
(See General Instructions - paragraph 5)

1. Address of customs post
2. Declaration that the waste has been exported from the customs territory of the Community
3. Date of exit
4. Date, stamp and signature of customs authority.

EUROPEAN COMMUNITIES

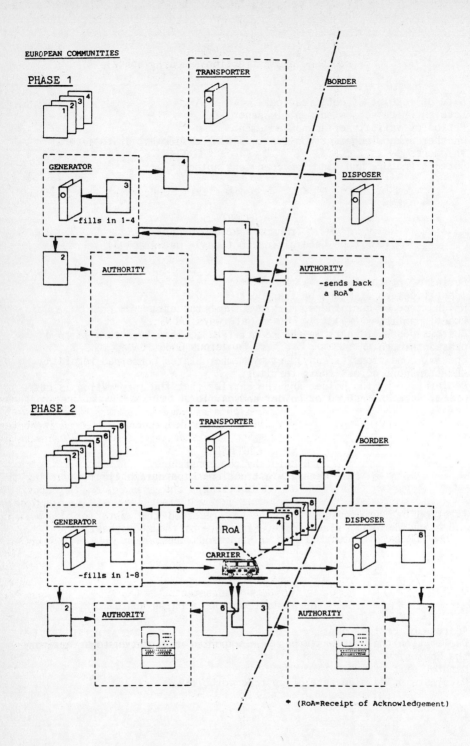

* (RoA=Receipt of Acknowledgement)

Annex 12

SURVEY OF INTERNATIONAL AGREEMENTS DEALING WITH TRANSPORT
OF DANGEROUS SUBSTANCES

A number of international agreements exist concerning the transport of dangerous substances:

-- International Regulations concerning the carriage of Dangerous Goods by Rail (RID);

-- European Agreement concerning the International Carriage of Dangerous Goods by Road (ADR);

-- European Agreement concerning the International Carriage of Dangerous Goods by Inland Waterways (ADN);

-- The International Maritime Dangerous Goods Code (IMDG).

<u>Road transport</u>

In 1958, a European agreement dealing with transport of dangerous goods by road (ADR) was concluded. By 1968, this agreement had been ratified by 5 countries and so came into force. The ADR provided to most affected countries a major basis for internal transport legislation governing movement of dangerous goods.

The ADR contains Annexes which set forth specifications regarding the substances, their packaging and labelling, and manifests (a copy is attached - CMR), as well as requirements for vehicles, construction, equipment and operation.

The ADR has divided the list of dangerous substances into eight classes:

class 1a   explosive substances and articles

class 1b   articles filled with explosive substances

class 1c   igniters, fireworks and similar goods

class 2   gases; compressed, liquified or dissolved under pressure

class 3   inflammable liquids

class 4.1   inflammable solids

class 4.2	substances liable to spontaneous combustion
class 4.3	substances which give off inflammable gases on contact with water
class 5.1	oxidizing substances
class 5.2	organic peroxides
class 6.1	toxic substances
class 6.2	repugnant substances and substances liable to cause infection
class 7	radioactive substances
class 8	corrosive substances.

Railway transport

In the 1920's, the RID was added as an annex to already existing conventions governing the transport of goods by rail (CIM). The format of this agreement is very similar to that of the ADR, in terms of the classes and substances since the RID was used as an example for the ADR. A copy of the CIM manifest is attached.

Waterway transport

There is a European agreement, ADN, governing inland waterway transport. In 1970, a separate agreement was concluded for the transport of dangerous goods on the Rhine, for which the ADN served as model.

Maritime transport is coordinated by the International Maritime Organization. An International Maritime Dangerous Goods Code (IMDG) was drafted which is based upon the Convention on Safety of Life at Sea (SOLAS) and also upon the UN class numbering and packaging and labelling regulations. The IMDG serves as the main basis for national legislation of many signatory countries. However, in cases involving water transport, there are no requirements for uniform transport document or manifests.

*

*   *

In summary, for a transport by road and by rail, requirements for a uniform manifest exist but not for water transport. Where required manifests always accompany the transport. However, no information flows (except for statistics) between shipper, consignee and governmental authorities. Legislation governing transport of dangerous goods does not require such information flows.

1

Exemplaire pour **expéditeur**
Exemplaar voor **afzender**
Exemplar für **Absender**

Code transporteur No
Vervoerderscode
Code Frachtführer Nr

1 Expéditeur (nom, adresse, pays) / Afzender (naam, adres, land)
Absender (Name, Anschrift, Land)

LETTRE DE VOITURE - DOCUMENT DE TRANSPORT
VRACHTBRIEF - VERVOERDOCUMENT
FRACHTBRIEF - TRANSPORTDOKUMENT

CMR NL 0190662

Ce transport est soumis, nonobstant toute clause contraire, à la Convention relative au contrat de transport international de marchandises par route (C.M.R.)

Dit transport is, ongeacht enig tegenstrijdig beding, onderworpen aan het Verdrag betreffende de overeenkomst tot internationaal vervoer van goederen over de weg (C.M.R.)

Diese Beförderung unterliegt trotz einer gegenteiligen Abmachung den Bestimmungen des Übereinkommens über den Beförderungsvertrag im internationalen Straßengüterverkehr (C.M.R.)

2 Destinataire (nom, adresse, pays) / Geadresseerde (naam, adres, land)
Empfänger (Name, Anschrift, Land)

16 Transporteur (nom, adresse, pays) / Vervoerder (naam, adres, land)
Frachtführer (Name, Anschrift, Land)

3 Lieu prévu pour la livraison de la marchandise (lieu, pays) / Plaats (bestemd) voor de aflevering der goederen (plaats, land) / Auslieferungsort des Gutes (Ort, Land)

17 Transporteurs successifs (nom, adresse, pays) / Opvolgende vervoerders (naam, adres, land)
Nachfolgende Frachtführer (Name, Anschrift, Land)

4 Lieu et date de la prise en charge de la marchandise (lieu, pays, date) / Plaats en dat. v. inontvangstneming der goederen (plaats, land, datum) / Ort und Tag der Übernahme des Gutes (Ort, Land, Datum)

18 Réserves et observations du transporteur / Voorbehoud en opmerkingen van de vervoerder
Vorbehalte und Bemerkungen des Frachtführers

5 Documents annexés / Bijgevoegde documenten
Beigefügte Dokumente

6 Marques et numéros / Merken en nummers
Kennzeichen und Nummern

7 Nombre de colis / Aantal colli
Anzahl der Packstücke

8 Mode d'emballage / Wijze van verpakking
Art der Verpackung

9 Nature de la marchandise / Aard der goederen
Bezeichnung des Gutes

10 No statistique / Statistisch nummer / Statistiknummer

11 Poids brut. kg / Bruto gewicht in kg / Bruttogewicht in kg

12 Cubage m3 / Volume in m3
Umfang in m3

Classe/Klas Klasse | Chiffre/Nummer Ziffer | Lettre/Letter Buchstabe | (ADR*)

13 Instructions de l'expéditeur / Instructies afzender
Anweisungen des Absenders

19 Conventions particulières / Speciale overeenkomsten
Besondere Vereinbarungen

20 A payer par / Te betalen door / Zu zahlen vom | Expéditeur / Afzender Absender | Monnaie / Geldsoort Währung | Destinataire / Geadresseerde Empfänger

Prix de transport / Vrachtprijs
Fracht:
Réductions / Kortingen
Ermäßigungen:
Solde / Saldo
Zwischensumme:
Suppléments / Supplementen
Zuschläge:
Frais accessoires / Bijkomende kosten / Nebengebühren: +
TOTAL / TOTAAL
GESAMTSUMME:

14 Prescriptions d'affranchissement / Frankeringsvoorschrift
Frachtzahlungsanweisungen
☐ Franco / Frei
☐ Non franco / Niet franco / Unfrei

21 Etabli à / Opgemaakt te
Ausgefertigt in | le / de am | 19

15 Remboursement / Rückerstattung

22 | 23 | 24 Marchandises reçues / Goederen ontvangen
Gut empfangen

Lieu / Plaats Ort | le / de am | 19

Uitgave Stichting Vervoeradres - Den Haag

Model IRU-Benelux 1976

143

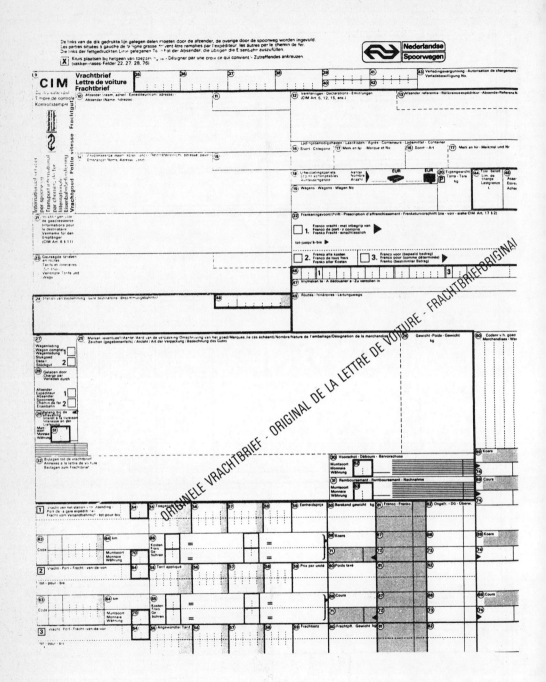

TOWARDS A MUTUALLY ACCEPTABLE INTERNATIONAL DOCUMENT TO MONITOR THE TRANSFRONTIER MOVEMENT OF HAZARDOUS WASTE

John Butlin
Consultant to Environment Directorate
OECD, Paris

SUMMARY

The international flow of hazardous wastes has become an environmental and economic fact of life. Estimates of the quantities transferred vary between countries, but Butlin and Lieben (1) estimated that up to 25 per cent of the wastes generated in OECD Europe and North America, are treated and/or disposed in a country other than that in which they arose. About three quarters of this is disposed at sea; the remainder is stored, treated and/or disposed on land in another country. Concern has been expressed for a number of years over the problems of international control over substances in these wastes which, if released into the environment in the form in which they are transported, could impose a serious immediate and perhaps long-term threat to the environment in the country for which they are destined or a transit country. They may also impose an immediate or long-term threat to human health.

The purpose of this paper is to examine the feasibility of using a trip ticket (2) to accompany transfrontier movements of hazardous waste (3). Implementing such a system is one way in which international control of these wastes can be achieved. This paper goes further and examines the feasibility of implementing an international control system using such a document.

The paper is in five sections: after the introduction, the problems raised specifically by the international dimension are reviewed. In the third section, the desirable characteristics of an international trip ticket are outlined. This section has in common with the fourth section a concern that an international trip ticket should be compatible with modern data processing systems, the fourth section being concerned with support systems for the document. The proposals of the European Commission for a Community-wide document, and of the United States for a federal trip ticket in place of the many State documents that had been developed, are considered in the fifth section. The final section summarizes the discussion and draws such conclusions as are to be found in it.

INTRODUCTION

In this section we review the purpose of the trip-ticket systems currently operating in Member countries; the question of notification of safe arrival or failure to arrive within a prescribed time period; and the responsibility of the public sector and private industry in the treatment, storage and disposal (TSD) system. It places the discussion in the context of trip-ticket systems currently in operation in Member countries.

1. The Purpose of a Trip-Ticket System

As noted above, the purpose of any trip-ticket system, whether national or international, is to facilitate control over the management of hazardous wastes. Most Member countries have decided that, to be effective, control must be exercised from the point and moment of generation to the point and moment of disposal (so called "cradle-to-grave" control). Some countries have also introduced regulations concerning the management of disposal sites up to, and following, their closure. This is quite another topic, however, and we are concerned only with the generation to disposal process in this paper.

We are concerned, then, with the control of wastes within the TSD system to ensure that they reach the desired and approved destination without undue delay, with minimum risk to the environment, to those involved in the TSD chain, and to the public in general. In addition, the system should not involve the generator of the waste in excessive additional costs.

In addition to facilitating cradle-to-grave control, the trip-ticket, if used properly, offers other advantages:

-- Wastes can be identified in case of accident (and appropriate remedial or preventive steps taken if necessary);

-- Under a pre-notification system there is some opportunity to plan for the disposal of certain categories and volumes of wastes;

-- It provides useful information on the quantities and types of wastes generated, and on the geographical distribution of these wastes. In terms of ensuring that sufficient capacity for hazardous waste management is available where it is required, this information is valuable.

2. Variations in National Systems in Existence

Not all Member countries of OECD currently operate a trip-ticket system as part of their hazardous waste management strategy. The countries that currently operate a national trip-ticket system include: Germany, the Netherlands, the United Kingdom and the United-States. Switzerland currently is in the process of introducing a system, and the European Community is developing proposals for a Community wide trip-ticket system. These systems differ in a number of ways:

-- Whether notification is given before or after shipment of the wastes; or whether it is part of the trip ticket process;

-- Whether the public authorities or the private parties involved are responsible for ensuring that the waste arrives at the proposed TSD site within a specified time limit;

-- Whether provision is made for regular shipments.

The Dutch system is the only one where documents concerning shipment are submitted following the shipment, although the generator must have received agreement from the proposed recipient facility before the waste is shipped. In all other countries which have a notification system, notification must be given and the documents sent either prior to the waste being dispatched, or the documents must accompany the waste.

The question of responsibility for tracing wastes that have not arrived at the intended destination within a certain period of time usually falls on the responsible public bodies, be they national as is the case in the Netherlands or regional, as is the case for example in Germany or the United Kingdom. In the case of the United States, however, it is the responsibility of the generator to ensure that the wastes have arrived at the intended destination. In the event of their failing to arrive, the generator can then inform the regional branch of the U.S. Environmental Protection Agency, but no public authority, either federal or state, is involved in the routine administration of the manifest system.

Some countries make provisions in their notification or manifest systems for regular shipments of specified wastes to made under an umbrella notification without requiring each consignment to be accompanied by a separate manifest. The Netherlands and the United Kingdom both have such provisions, for example. However, anecdotal evidence from a few UK Waste Disposal Authorities suggests that only a small proportion of those who are entitled to take advantage of this provision actually do so. Such a provision does, of course, raise the question of what variations in composition of the wastes are to be tolerated. In the EEC Directive on Transfrontier Shipment there is a provision for regular shipments of wastes of similar composition.

PROBLEMS RAISED BY THE INTERNATIONAL DIMENSION

Although there is a clear similarity between the national systems currently operating and those proposed (for example, in Austria, Canada, Switzerland and Sweden) the transport of wastes across national frontiers raises a number of specific problems unique to the international problem. For example, are national systems of hazardous waste classification compatible? What is the role of the customs authorities in the international waste management system? Are problems raised due to incompatibilities between existing national and proposed international systems for managing the flow of hazardous wastes from point of generation to point of ultimate disposal? We shall consider each of these in turn.

1. Compatibility of National Systems

H. Yakowitz (4) has developed a cross-reference system for the hazardous waste classification systems currently operating in several countries. As a result of developing this system he has come to a number of interesting conclusions. Amongst those relevant to our current purpose are the following:

-- "There is no internationally accepted definition of which properties and/or provenance of a substance designates that specific substance to be a hazardous waste";

-- "(Discrepancies between national classification systems) are likely to create difficulties since all transfrontier consignments must comply with local legislation and regulations in the country of origin, transit countries and the (country for which the wastes are destined)";

-- "Monitoring authorities e.g. customs officials require some means of assurance that consignments comply with local requirements. Furthermore, this means of assurance must be rapid, comprehensible and as low in cost as practical since many shipments may occur";

-- "Some international lists of proscribed wastes exist including that in the draft EEC Council Directive and wastes proscribed from sea dumping by means of five international conventions";

-- "The United Nations recommended classification system, which deals with transport of all dangerous substances, can be applied to some extent to existing lists of hazardous wastes".

-- "Existing lists of hazardous wastes for several OECD Member countries can be ...incorporated into a fairly simple, comprehensive cross-referencing system. The international list can be... included in the system";

-- "The cross-referencing system can be used in conjunction with a simple transport document to provide (the basis for) adequate control over transfrontier movements of hazardous wastes.

From this work it is clear that the cross-reference system is a useful tool in the transfrontier management of hazardous wastes. Other, perhaps more simple, management instruments are also likely to be of use in the practical operation of such a system. Key amongst these is a glossary, or dictionary, of the wastes concerned in a number of Member country languages (5). Similarly, a document summarizing the customs requirements and national hazardous waste management regulations (including specific insurance requirements for hazardous waste transport, and any liability which is specifically directed towards the generator or transporter of the wastes). The work of F. van Veen (6) and J.P. Hannequart (7), presented at this Seminar go some way towards this, but would need to be integrated, reduced in scope, include additional material on customs procedures and be more "user friendly" (8).

From the discussion above it appears that differences between national systems need not prove an insuperable barrier to the establishment of an international manifest system as a contribution to the effective management of international flows of hazardous wastes. There are, however, two further issues which relate not to the possibility of establishing a system but to the likelihood of effective implementation and control of such a system. Specifically, the role of the customs authorities should be investigated, in particular the degree of responsibility to be placed on these border authorities.

2. The Role of Customs Authorities

It is, in principle, possible to conceive of customs authorities having a spectrum of responsibility for enforcing regulations concerning transfrontier flows of hazardous wastes, ranging from the purely formal requirement of ensuring that the appropriate documents are being carried, and that these have been properly completed (although this would immediately raise the question of different languages), to giving customs authorities the right to verify that the constituents of the waste are as specified on the international transport document. Although this could involve delays in the onward transport of the wastes, it in no way exceeds powers that customs authorities in Member countries have at the moment. A number of unresolved questions are raised in the latter case, however:

1. Who will bear the cost of the sometimes timely and often expensive procedures involved? (including how can these costs be recovered?);

2. What procedures are to be followed in the event of wastes not being accepted by the Stated consignee?

3. Should wastes intended for recycling be subject to such a procedure?

With respect to the first item, the Polluter-Pays-Principle is the criterion by which the costs associated with the management of hazardous wastes are allocated, including such "frontier" costs as analysis, either for random spots checks, or in the event of the information on the consignment note being suspect. However, the problem arises not in the principle to be applied, but in the application of this principle when the wastes are being handled by a transportation enterprise, or when they are in another country and therefore subject to a different national jurisdiction. In other words, the problem arises with the recovery rather than the allocation of these costs in the transfrontier shipment of the wastes.

The procedures to be followed in the event of a consignment being rejected are again, in principle, simple. It can be made a condition of the contract that if wastes do not conform with the packaging, transport conditions or analysis specified in the contract they will be accepted by the consignor (such is the case with the Herfa-Neurode underground disposal operation, for example). However, the problem arises in the event of consignments of particularly toxic wastes where the information that has been provided to the responsible authorities in the country where the wastes arose is incorrect, or where a transit country, under similar circumstances, refuses to permit return transit of such wastes.

The question of whether wastes for recycling should be included in or exempted from a transfrontier flow of waste control system is particularly complex. We will not address here the question of whether inclusion would hinder the recycling effort. We are more concerned with the question of whether a derogation for materials for recycling would prove too easy a loophole through which transfrontier shipments of wastes ultimately intended for disposal could pass. Arguments have been advanced both for and against this; we believe that pre-shipment acceptance of the wastes by an approved recycling facility, and a burden of proof falling on the recycling enterprise to demonstrate to the competent national authorities that the wastes have, in fact, gone through a materials recovery process, would minimise the abuse of a recyclying derogation.

A further issue which could be raised in relation to the transfrontier flow of hazardous wastes concerns the possibility of restricting the transit points between any two countries to a limited number at which there are special facilities for analysing the contents of hazardous waste consignments, and at which customs officers are stationed who have received some specialist training are based. Clearly, there would again need to be some agreement on cost-sharing for these analyses between the countries concerned (or an agreement on who would bear the costs); agreement to designate the same acceptable crossing points between contiguous countries would also appear to be a pre-requisite for the successful implementation of such a system.

DESIRABLE CHARACTERISTICS OF AN INTERNATIONAL TRIP-TICKET

An international trip-ticket needs to display at least five desirable characteristics:

1. It should be as simple as possible whilst still fulfilling its purpose of being the keystone of effective international hazardous waste management;

2. It should use as few languages as possible;

3. It should contain as few copies as possible;

4. It should be compatible with existing customs procedures (or require as little modification of existing procedures as possible);

5. It should be compatible with modern data processing systems.

Whilst some of these "ideal" characteristics may seem mutually incompatible, it is worth our while to consider each in a little more detail. The virtue of simplicity is self-explanatory: any document which is going to be acceptable to 10, or 24 (or more?) countries can only ask for the most elementary information. This, however, begs the fundamental question: "What is the minimum amount of information which can be required whilst maintaining the effectiveness of the system?" Views on this obviously differ, but the following suggestions are made to stimulate discussion. It is suggested that the waste be described by its chemical analysis, and classified according to

150

the appropriate code/classification, in the country of origin, any transit country(ies) and the country of final destination in addition to basic information on the generator, transporter and disposer. The language in which the form is completed should be one among a few languages (see below), and the codes those issued by the competent authorities in the individual countries as amended from time to time. This approach has several advantages: it is simple, the analysis is verifiable, and it requires only that imported wastes (whether for re-export or final disposal) be classified in the same way as domestic hazardous wastes in that country.

The number of languages acceptable for use in an international trip-ticket to some extent depends on the countries involved, and to some extent on the international organisations involved in developing the system. The number of languages acceptable for an international trip-ticket is a combination of convenience and logistics. The more acceptable languages there are the more convenient is the system for the individual countries concerned, but the greater the amount of cross-referencing required, and the more voluminous and costly the manuals needed to support such a system.

The number of copies to be included in an international trip-ticket is much more difficult to determine. The EEC proposed system uses an 8 copy manifest, together with a four copy notification document. It seems, in principle at least, possible to make the 8 copy document serve both purposes, and to forego the notification document. (In effect, the generator would need to receive an acknowledgement of receipt of the copy sent to the importing and transit countries before the wastes would be accepted either for transit or import. For a brief summary of the EEC system see report of van Veen (6).

Customs procedures are a specialised topic in themselves, and encompass broader issues than we are discussing here. We should note, however, that such a system, to be sucessful, would need agreement to be reached on the customs' procedures to be followed in dealing with shipments of hazardous wastes under the trip-ticket. The advantage of any system being compatible with electronic data processing can easily be explained: the purpose of the trip-ticket is to provide information with which co-operating countries can manage international flows of hazardous waste. The most efficient way to store, transmit and receive such information is via electronic means. A system such as that described is eminently suited to electronic data management systems, and would enable a degree of control, advice/warning of non-receipt/failure to transit that is much more simple and speedy than existing postal, telephone or telegraphic bilateral communication.

SYSTEMS NECESSARY TO SUPPORT THE INTERNATIONAL TRIP-TICKET

To some extent this topic has already been covered. We noted above that a cross-referencing system by itself would probably be insufficient (although clearly necessary), and that other, supporting materials, such as a handbook, a glossary of hazardous wastes in different languages, and a manual of agreed customs procedures, special insurance and generator/transporter liability provisions would also be necessary. We have also noted the efficiency of being able to transmit to all co-operating countries, in all agreed languages, the status of any proposed or ongoing consignment of hazardous wastes. There are, however, other more formal supporting systems. The first of these is the OECD Council Decision and Recommendation on Transfrontier Movements of Hazardous Waste, in addition to the activities of other international organizations such as the United Nations Economic Council for Europe and the United Nations Environment Programme. The latter is developing guidelines on the environmentally sound management of hazardous wastes, and the former already has in place its Transport of Dangerous Goods Recommendations with a working group seeking to extend the provisons of these to hazardous wastes. It is also clearly necessary that national monitoring and enforcement systems be in place and operating in all participating countries, as it is ultimately the effectiveness of these systems (and the responsible authorities who implement them) which will determine the success or failure of the international trip-ticket. It may also facilitate the operation of such a system if each participating country were to designate the sites within its jurisdiction at which imports of particular categories of wastes could in principle be accepted (without prejudicing the right of a particular facility to accept or reject a particular waste at a particular time). It may appear that such a procedure, authorizing or designating certain facilities as centres to which hazardous waste imports could be permitted. Whilst we are not proposing licensing as a necessary support system, there are clearly advantages in terms of monitoring and enforcement if the number of facilities at which hazardous waste imports can be treated or destroyed is limited.

RELEVANT EXPERIENCE IN OECD MEMBER COUNTRIES

There are two sets of experience in Member countries relating to the implementation of either transborder or international trip-ticket systems, although neither are yet operational (9). The form developed by US Environment Protection Agency is a relatively simple document, which uses only four copies, and which incorporates a small quantity of supplementary information which may be required by individual states but is not required by the federal authorities. In the information proposed above for an international manifest this information is provided implicitly through the information contained in national codes or classification systems. The number of copies required for most countries would be greater than four, because of the need to inform the responsible authorities.

The EEC Directive on the Supervision and Control Within the European Community of The Transfrontier Shipment of Hazardous Waste (10) incorporates proposals both for a notification document and a consignment or trip ticket document, as noted above. The system, which is to be implemented by all Member States by October 1st, 1985, will operate in all of the Community languages. The trip ticket itself* incorporates similar information to the national manifest in the United States, but, in addition, requires information concerning:

-- the point of generation of the wastes;
-- the nature of the transporter's third-party insurance;
-- the process from which the waste originated.

152

The major problem associated with the EEC manifest is the large number of languages in which it is to operate and the associated increase in paper work involved in importing/exporting hazardous wastes with an 8 copy form. The use of Community-wide electronic systems to communicate this information would facilitate the process, although there are no plans at the moment to use electronic rather than written means of communication for this systems.

Whilst neither of these systems is operating yet, they both demonstrate in their draft form that it is possible to design a trip-ticket (to be completed in one language only) which is simple, and would easily be managed with a simple electronic data processing system. Completion of either form by an experienced clerk furnished with the correct information would be a matter of 10-15 minutes, and the regular consignments' provision of the EEC system would reduce this even further. In comparison with either of these forms the information proposed above for a more general international trip-ticket is slightly greater, but in effect only requires the same information as that required in the United States, except that it must be provided according to the classification schemes of the generating, transit and importing countries. It is in the light of this that the importance of appropriate supporting systems can most clearly be seen.

CONCLUSIONS

This paper has outlined the issues relating to the development of an international trip-ticket as the keystone of an international management sytem for the movement of hazardous wastes across national borders. The prime reason for adopting such a trip-ticket is to effect control from point of generation to point of disposal (so-called "cradle to grave" control). A number of OECD Member countries currently have a trip-ticket/manifest/consignment note system in place, and several more are considering or developing systems. While there is a degree of variation between individual national systems, the prime motivation is always to achieve more effective control, although there are other advantages to the system.

With national systems increasing rapidly in number, there is a clear logic in extending the system to movements of hazardous wastes between countries. The international dimension brings about a number of problems, however, of which the major one is the different systems used by individual countries to classify wastes as hazardous or not. Both this problem and other problems specific to the international movement of hazardous wastes could be overcome with the development of documents already available or, in the case of the multibilingual glossary of hazardous waste terms, yet to be developed. Any international trip ticket should display a number of characteristics, of which simplicity is clearly the most important. Any trip-ticket system operating at an international level needs both formal and operating support systems, of which the most important is an operational and effective monitoring and enforcement system in each participating country.

Both the United States and the European Commission have developed draft systems for the transfrontier (state and national respectively) movement of hazardous wastes. In the body of this paper it is proposed that an international trip-ticket should bear information on:

-- the description of the waste(s) in one acceptable language agreed upon by all parties;

-- the relevant classification codes for the country in which the waste arose, through which it was transported, and in which it was treated, stored or disposed;

In addition to basic information concerning the generator, transporter(s) and disposer. Such a trip-ticket system would most effectively be managed using modern techniques of electronic data processing, the speed and efficiency of which would greatly enhance the scope and efficiency of the control system.

It is clear from the work of OECD and other international organizations that an international trip-ticket would, in principle, be an effective tool in managing the international movement of hazardous wastes. Presuming that such a system were adopted amongst all OECD Member countries, much of the preparatory work for the supporting documentation has been initiated. There would, in addition, be considerable advantage to there being a uniform approach to the international movements of hazardous wastes amongst the OECD Member countries, particularly when discussing possible movements of these wastes to other country groupings.

NOTES AND REFERENCES

1. Butlin, J.A. and P. Lieben: "Economic and Policy Aspects of Hazardous Waste Management", Industry and Environment (Special Issue) 40.4, 1983, pp. 11-14.

2. Alternatively called a manifest, way bill, trip ticket or bordereau. We will use the term "trip ticket" in the paper.

3. Whilst accepting that terminology differs between OECD Member countries (describing these wastes alternatively as "special" or "dangerous") we retain the widely accepted adjective "hazardous".

4. See Report by H. Yakowitz in this book.

5. OECD has provided a most useful service in quite a different context. Its Agriculture Directorate has provided a 6 language dictionary of common names for fish, which is widely used throughout the world.

6. See Report by F. van Veen in this book.

7. See Report by J.P. Hannequart in this book.

8. It is only fair to indicate that neither of these reports were drafted with this purpose in mind. They serve, however, as an excellent base from which to develop it.

9. The US system is in a transition phase; it will be fully operational by September 1984. Forms are given in the report by F. van Veen.

10. Council of the European Communities: "Directive on the Supervision and Control within the European Community of the Transfrontier Shipment of Hazardous Waste, O.J.E.C., L 326 (13th Dec. 1984).

THE OPERATION OF A WASTE DISPOSAL FACILITY
WHICH ACCEPTS FOREIGN HAZARDOUS WASTE

Gunnar Johnsson
Untertage-Deponie Herfa-Neurode (UTD)
der Kali und Salz AG - Kassel
Germany

SUMMARY

The Underground Waste Disposal Facility (UTD) located at Herfa- Neurode has accepted hazardous waste from foreign countries since June 1974. This facility was able to take advantage of the unique geological properties and rock mechanics of the salt deposits of the Werra-Basin. The space remaining within the deposit following the mining operation in the potash-seams could thus be used to establish a disposal facility meant to deal with solid hazardous wastes. The procedure which has evolved for accepting hazardous wastes at UTD is described. The basic principles UTD personnel consider to be necessary for any transfrontier movement of hazardous waste are outlined as well.

INTRODUCTION

Since 1950 industrial production in OECD Member countries has increased steadily; production of chemicals has grown in real terms by about 3 per cent per annum. Industry, the public and responsible government entities realised that such growth would inevitably give rise to increasing amounts of waste. A certain amount of this waste might, in fact, be hazardous to man and the environment. Thus, a number of countries enacted legislation meant to control such wastes. Each country having legislation has compiled a list of wastes deemed to be hazardous (N.B. no two lists are the same; there is great variability in defining what constitutes a hazardous waste). The large amounts of hazardous waste being generated are viewed as a serious problem to man and the environment, not only in the industrial countries but in developing countries as well.

Certain events associated with inappropriate past disposal of hazardous wastes resulted in strong public interest in assuring that potential adverse

156

health, environmental and economic effects arising from hazardous waste management would be minimised. Hence, certain countries enacted special legislation. With respect to management of hazardous wastes, close international cooperation in industrial production and trade indicates that similar international cooperation and subsequent development of rules and regulations with respect to solving problems created by hazardous wastes is desirable, possible and necessary.

The capacity for treating, storing or disposing of various hazardous wastes (where their generation cannot be avoided) differs from country-to-country, for example, as a result of differing geological or technological conditions. Thus, there exists motivation for transfrontier movement of hazardous wastes in order to obtain environmentally sound elimination (recycling, incineration or disposal) of such wastes.

A large amount of time and effort has been invested by national and international organisations in order to try to cope with the challenges posed by hazardous wastes. The OECD Decision and Recommendation on Transfrontier Movements of Hazardous Waste, adopted on 1st February 1984, has set forth principles which can serve as the basis for future legislative action with respect to control of international shipments of hazardous wastes.

Accordingly, the major objective of this report is to document the experience of one facility, which since 1974 has accepted and disposed selected hazardous wastes imported for this purpose from foreign countries. The experiences of this facility may well contribute to the knowledge required in order to draft and implement practical international regulations. Public acceptance with respect to hazardous waste management activities in general, as well as the performance in "getting the job done" of waste generators, carriers and waste disposers, depends on the appropriate application of rules and regulations. At all stages, hazardous waste management activities must be open to public and governmental scrutiny.

THE UNDERGROUND WASTE DISPOSAL LOCATED AT HERFA-NEURODE (UTD)

The UTD was established in 1972 by Kali und Salz AG, Kassel, in compliance with existing waste legislation of the Federal Republic of Germany.

Extensive knowledge of the geological conditions in the local area -- supported by 80 years of mining experience -- provides the basis for the statement that wastes, deposited in the mined-out section of Herfa-Neurode potash mine are effectively withdrawn from the biocycle for geological periods. Without human intervention, such wastes can never enter the biosphere again. Figure 1 contains a representation of the basic geological situation and shows part of the empty mineworkings in the lower potash-seam.

The UTD installation is an integral part of the waste disposal plan of the State of Hessen but is available for certain wastes from all States of Germany. In the context of the german waste policy programme, the UTD helps to deal with a substantial challenge regarding protection of the environment. Recently, the underground waste disposal facility has also assumed growing importance for other European countries.

The existence of this facility provides waste generators with a stable source for environmentally appropriate disposal of hazardous wastes for the foreseeable future. If the operation of the UTD is carefully managed in terms of types and amounts of waste admitted, operations can continue for as long as the facility is needed.

FRAMEWORK FOR EXPORT-IMPORT OF HAZARDOUS WASTES

With respect to the acceptance of waste generated outside Germany, the procedure which must be implemented before a decision can be made as to whether a specific foreign waste will be accepted is essentially the same as for domestic waste. However, an additional import licence must be obtained for the waste. Figure 2 contains a list of the required steps in order to dispose of foreign waste at the UTD. Completion of each of these steps is absolutely necessary. Note that the adopted procedure includes adequate and prompt delivery of information to the authorities in both the exporting and importing countries.

Each of the steps outlined in Figure 2 will be discussed and explained in detail in what follows. The chances for mismanagement of hazardous wastes tend to occur when insufficient attention is paid to the detailed implementation of each step. Everyone concerned with the development of rules and regulations must clearly understand this particular point if appropriate and workable procedures governing transfrontier movement of hazardous wastes are ever to be fully implemented.

1.   Waste Generator: We presume that a generator outside Germany will have completed a careful survey and examination of domestic (local) and alternative methods for disposal of his hazardous wastes. Should such a generator then decide to seek permission to utilise the UTD, he submits an informal inquiry which contains an identification of the waste and the processes whereby the waste is generated. Furthermore, the waste must be characterised by means of an assessment of analytical data and physical and chemical properties pertaining to the waste.

2.   After a first evaluation of the submitted information, there is an inspection of the facility and processes of the waste generator by UTD-personnel. The UTD team evaluates the relevant analytical data and decides if additional investigation and documentation is necessary. The UTD team also specifies the technical conditions for disposal, packaging etc. of the waste in question. The waste generator designates those persons in his employ who will be responsible for the particular operation.

3.   If the waste generator finally decides to apply for permission to dispose in the UTD, he must request his proper governmental authority to provide official "confirmation" ("attestation") for the data given in the "Formblatt A" (see Annexes A.7 and A 3.1-A 3.2). By providing such confirmation ("attestation"), the governmental authority formally supports the waste generator's disposal petition. (This is developed and required by the

State of Hessen.) As a result, governmental authorities in both the exporting country and Germany are fully informed with respect to the intentions of the generator, the specific waste characteristics and the reason why that waste is to be exported for ultimate disposal in the UTD.

4.   Inquiry Form A (Formblatt A) together with the "confirmation" ("attestation") of the responsible authority is then submitted to UTD-Kassel for final examination. (If necessary, any additional information requested by UTD concerning the particular waste must also be supplied at this time).

5.   Careful examination of the disposal request is then carried out by personnel of UTD.   If UTD then confirms that the specific waste in question can be appropriately disposed underground, UTD submits "Formblatt A" (identified by a code number) to the Mining Authority responsible for the underground operation (Bergamt Bad Hersfeld).   Copies of the "confirmation" are also submitted to the Mining Authority for review.   After reviewing the disposal request data, the Mining Authority forwards a copy of the documents to the Environmental Agency of the State of Hessen for final evaluation with regard to hazardous waste management policy.

      As a result of these actions, all governmental authorities having an interest are informed.   In particular, documentation is available to confirm that the government authorities in the exporting country are officially supporting the intentions of the generator and are confirming the necessity and appropriateness of the proposed disposal action.   The properties, characteristics and quantities for the particular case are thus fully known to all parties of interest.

6.   If informal approval for the disposal is obtained from the Environmental Agency, the Mining Authority (Bergamt Bad Hersfeld) then examines the submitted data with particular attention to all aspects of the safety of the underground operations of the UTD.   If a decision to proceed with the disposal is finally made, the Mining Authority sends the final approved (signed) Form A (Formblatt A) to the management of UTD.   Copies of this Form A are provided to the regional authority and to the authority responsible for traffic control in the region.

7.   The UTD then executes a "letter of acceptance" (see "Schriftliche Annahmebestätigung", Annex A.2) which is transmitted to the waste generator. A copy of that letter is provided to the carrier.   Note that the transporter must be mutually agreed upon by both the waste generator and UTD.

8.   At this point, the transporter can apply for the import licence at the District Authority which is responsible for the UTD (Regierungspräsident Kassel).   This authority keeps the Ministry of the Environment of the State of Hessen informed by sending a copy of these documents.   If, after carefully reviewing the political implications associated with the prospective waste import, there are no objections, the Regierungspräsident Kassel will execute the import licence.   This licence is then forwarded to the transporter.   The documents needed by the generator and transporter are: "letter of acceptance

of the UTD together with Form A (Formblatt A) as signed by the Mining
Authority, "General Business Condition" ("Allgemeine Geschäftsbedingungen"
UTD), the usual form utilised for import of waste (§ 13 Abfallbeseitigungesetz
der Bundesrepublik) (see Annexes A 2., A 3.1, A 3.2, A.4 and 6.1-6.4)) and the
"confirmation" (see point 3 above).

9.   The carrier provides copies of the import licence to the UTD which then
forwards this licence along with the original "Letter of Acceptance" to the
waste generator.   The waste generator may now begin to dispatch that
particular waste to the UTD.   The "Letter of acceptance" of the UTD for this
particular waste remains valid for an indefinite time.   The corresponding
import licence is normally valid for a specified quantity of waste for three
years.   After three years the import licence may be extended for one
additional three year period.

      In the opinion of UTD, this set of steps constitutes an indispensable
procedure without which no fully controlled transfrontier movement of
hazardous waste can be performed.   Having completed each step, the waste
generator and the operator of the disposal facility are in a position to
justify, explain and, if necessary, to "defend" the actions related to a
particular case.   All parties concerned or involved in the decision to permit
the waste to be transported to the UTD thus have the basis for a sound and
reliable public information mechanism.   Such a mechanism is a key prerequisite
for public acceptance of appropriate transfrontier movement of hazardous
wastes for ultimate diposal in another country.

      In the absence of public knowledge of activities associated with
hazardous waste disposal, public opposition may well occur.   In such cases
public media may "pour oil on the fire" so that applying rational and
reasonable solutions to what is, in fact, a fairly well defined technical
problem, is effectively precluded.   Indeed, in such cases, politicians,
administrators, the waste generator and waste disposal operator may be forced
to react irrationally.   The events associated with the so-called Seveso affair
represent only one example of the possible consequences of such a situation.

Summary of steps 1 to 9 above

      The steps discussed above deal with essential technical decisions
concerning the disposition of hazardous wastes.   Such technical decisions are
first and foremost the responsibility of the waste generator and the waste
disposal facility.   These entities must have the knowledge and the
responsibility for evaluating and determining the basis on which to take
certain actions and ultimately to demonstrate and to make plausible to all
concerned what has happened and what will happen to a given waste, from
"cradle to grave".   And the generator and his chosen disposal firm and mode
are first and last responsible and liable for the consequences of activities
with respect to a given waste.

      This responsibility and liability suggests that no third party (for
example "waste broker" or other mercantile agent or dealer) should ever be
involved in the process.   Such parties never can be provided with the
appropriate and adequate knowledge and the responsibilty and liability for the
waste.   Many of the unfortunate "incidents" involving hazardous wastes in the

past occurred mainly as a result of such third parties making their "contribution".

Length of time needed for a foreign waste generator to obtain the import licence

Steps 1 to 3 are UTD "in house" jobs; the time needed to satisfactorily complete the required work strongly depends on the particular case and on the specific situation. The tasks may require one week or several months.

Steps 4 to 9 require the respective authorities to be engaged. The average time needed to accomplish these steps is about ten weeks, mainly because paper has to be sent hither and yon. In cases of emergency, when rapid action is absolutely necessary, the whole procedure can be completed in one week or even less. However, this time reduction requires that the case be put forward in person by the waste generator and representatives of the waste disposal facility who must present and defend all relevant documentation to each respective governmental authority. In some cases, a preliminary decision can be reached following direct contact between the authorities of the exporting and importing countries.

Conclusions regarding the framework for export-import of hazardous wastes

In ten years of experience over 100 000 t of foreign waste have been disposed in the UTD; this figure represents nearly 30 per cent of all wastes disposed in the UTD (see Figure 3). Thus, the framework governing transfrontier movement of hazardous wastes to the UTD has become a routine procedure. Persons engaged in using the framework are experienced and educated in its implementation on both sides of the border. Not only representatives and responsible personnel of the waste generator must visit the UTD prior to contracts being executed, but also appropriate officials of the government of the waste generator's country, i.e. members of the national organisations in charge of, or concerned with, hazardous waste management in the generator's country. In turn, the management of the UTD must ensure that UTD personnel have adequate knowledge of the national legislation and organisation in the respective countries with respect to hazardous waste management policy. These interactions are indispensable in assuring proper approaches to transfrontier movement of hazardous wastes. Adhering to the framework does not require undue administrative effort if the competent authorities are clearly identified and duly qualified personnel are made available.

Transfrontier movement of hazardous wastes is, in many respects, a politically and economically sensitive issue. In terms of existing regulations (national or international) with respect to sound environmental protection policies, disposal activities involving transfrontier movement of hazardous wastes can only be justified if an improvement in environmental protection is obtained for all parties concerned. The principles of a free market simply cannot be applied without restriction to the movement of hazardous wastes. Indeed, the economic and commercial aspects are of secondary significance; protection of the environment against pollution by hazardous wastes must be the primary goal of all actions and of all parties involved with transfrontier movement of hazardous wastes.

TRANSPORT

In order to provide a clearer description, transport operations will be described as if divided into two parts; typical transfrontier transport procedures and additional requirements with respect to hazardous wastes will thus be treated separately. Responsibility for safe transport rests with the waste generator and his chosen carrier. After reaching agreement with respect to the date of delivery of the waste with UTD, the waste generator arranges final details with the carrier.

1. Normal formalities regarding transfrontier transport

Denmark

The waste generator provides to the carrier: (1) Export-declaration (three-fold), (2) Custom-Invoice, or "pro-forma factura", and (3) T.2. Form (Angivelse til faelleskabs foersendelse, Déclaration de transit communautaire) (Versandschein). The carrier should already have obtained the usual Consignment-note (way-bill) (four-fold) and a copy of the waste import licence.

At the border, Danish Custom officials validate the T.2. Form and retain the export declaration. The driver receives a tracking-form (control-tag). German Custom officials retain the tracking-form and the validated T.2. Form containing the Import-Endorsement. The "pro forma factura" is validated and one copy is kept by customs officials. The carrier remits the Import value-added-tax on behalf of the waste generator; this tax is computed with respect to the freight haulage price, as shown in the Consignment Note. (This payment is subsequently refunded upon proper application on the part of the waste generator, by the Finance Administration of Germany).

France

The waste generator must bring the T.2. Form (Déclaration de Transit Communautaire) (four-fold) the "pro-forma factura" and the Export declaration to the local Custom Office having jurisdiction. There the T.2. Form is "opened" and validated. One copy of "T.2." is retained together with the export declaration. The carrier receives three validated T.2. Forms as well as the validated "pro forma factura".

At the border, the French Custom officials retain one T.2. Form and give the driver two validated T.2. and a tracking form. German Custom officials validate the "pro forma factura" and retain two T.2. Forms, the tracking form and "pro forma factura". (Tax payment as for Denmark). One T.2. Form with a validated import endorsement is returned to the French Custom officials in order to "close" this "T.2. file".

For the Netherlands and Belgium, the procedure is virtually the same as outlined above. For Switzerland and Sweden, some minor differences occur because they are not member countries of the European Communities.

2. Implementation of additional procedures pertaining to transfrontier movement of hazardous waste

Upon collecting the waste consignment, the carrier completes the required German Waste Trip Tickets or manifests (Abfallbegleitschein, see Annex) (six-fold) by inserting the number of pallets, drums and the respective German five digit code number designating the waste. The manifest must be signed by a responsible agent of the waste generator. The driver signs the manifest and adds the date. The driver also enters the number of pallets, drums, code numbers and designation of the waste onto the normal consignment note (way bill) so that these numbers correspond with the waste manifest.

At the border, custom officials of the exporting country implement normal procedures, by comparing Consignment note and "pro forma factura" and "export declaration", in which the nature and quantity of the load is shown. German Customs officials do the same. But in addition German Customs officials take possession of the waste manifest in order to compare the information therein with the Import Licence. (§ 13 German Waste Law) which the driver must also tender, following inspection of the load. German Custom officials enter onto one copy of the Import Licence (which is retained at this customs office), the amount of waste imported and the date. Next, the quantity of such waste imported to date into the Germany is compared with the amount permitted by the particular Import Licence.

If the Import Licence is no longer valid, either by means of expiration of the licensed period, or by means of exceeding the quantity of waste licensed, the Import Licence is returned to the governmental entity which issued the licence, along with documentation of the dates and the quantity of the specific waste imported. Thus, comparison with such return waste manifests is easily achieved if needed. Close control of both the carrier and waste disposal facility can be maintained, and the "waste streams" can be evaluated in detail by Germany. Moreover, the yearly reports of the UTD, which record the wastes imported by country, waste generator and carrier for each waste can be consulted. As a consequence of recent improvements in data processing, the UTD figures for each waste code number, identifying the waste and the waste generator and, if required, the carrier and the date of arrival for each load, are easily accessible. After the waste load arrives at the UTD, and undergoes and passes inspection and control procedures, the waste manifest (six-fold) is signed by the personnel of the UTD, to document that the waste was accepted.

Waste manifest copies Nos. 1, 2 and 3 are given to the carrier, who in turn gives No. 1 and the freight invoice to the waste generator. Copy No. 2 is provided to the German agency which issued the Import Licence. (For the past several years the carrier has been required to give an extra Copy No. 2 to the foreign authority responsible for the waste generator and which has provided the "confirmation" mentioned in 2.3 above. UTD personnel felt it was necessary to "close" this circuit as is done in Germany).

No. 3 is retained by the carrier and becomes part of his documentation.
(§ 11, German Waste Disposal Law, Abfall Nachweisbuch).

The UTD retains Copies No. 4, 5 and 6 of the waste manifest. Copy
No. 4 is sent to the agency which issued the import licence. Copy No. 5 is
sent to the foreign waste generator, along with the monthly invoice requesting
payment for disposal services. The waste generator is thus able to inform the
responsible authorities in his country that the particular waste has been
eliminated as planned. Copy No. 6 remains at the UTD and becomes part of that
facility's documentation.

3. Transport by railroad

About 25 per cent of the waste arriving at the UTD is transported by
rail; some of this material is shipped by foreign waste generators, if rail
transport is expedient. In this case, the entire Transfrontier-Procedure
described in sections 1 and 2 above takes place at the Station of Bebra, close
to the UTD, where a local Customs office is located. The implementation of
the procedure is accomplished by a forwarding agent mutually agreed upon by
the waste generator and UTD. The waste manifests are completed at the station
after control and inspection of the load at the station. The import licence
is deposited at the local Customs office in Bebra.

4. Application of the regulations for transport of Hazardous Goods (ADR)

Some hazardous wastes are, with respect to the stipulations of the ADR,
"Hazardous Goods". Waste loads must be "assimilated" into the ADR system by
providing adequate code numbers and adhering to the articles of whatever
regulation is applicable. In some cases, this procedure is not unequivocal.
But for packaging of the wastes, adherence to ADR rules is essential. In this
case, it is advisable to refer to the most hazardous component. Because the
UTD accepts only solid wastes -- those not producing inflammable gases and not
self-ignitable -- in some cases, more stringent packing may be done than is
strictly necessary.

UTD personnel recommend that ADR rules for hazardous waste be amended
in order to avoid misleading declarations, designations, packaging and
labelling. There should be a special marker on trucks transporting "Hazardous
Waste", in addition to the orange plate. A carrier of hazardous waste
obviously must have all applicable safety instructions on board for the
particular waste being transported. Improvements in the adaptation of
existing regulations are needed so as to avoid any shortcomings with respect
to safety aspects in case of traffic accidents and/or possible spillage. In
the absence of sensible regulations, governmental authorities and persons who
must control traffic in hazardous wastes might be faced with impossible tasks
if designation and labelling is not clear-cut for such wastes. In any case,
the carrier has both the import licence and "Form A" (Formblatt A) plus the
"letter of acceptance" of the UTD on board, all of which describe the waste
and its properties. On those forms persons are named who can provide further
detailed information, either from the waste generator or the UTD.

5. Conclusions relating to points 1 to 4 above

The text above described the vital steps which must be performed in order to accomplish transfrontier movement of hazardous waste. In addition to the normal procedures governing transfrontier traffic of goods, only a few special documents are required in order to achieve reliable control and monitoring of transfrontier movement of hazardous wastes. The tools are available. As in all human activities, rules and regulations can be violated, either by negligence or by intention. But even if violation occurs, the procedures described, combined with the quality of institutions and persons involved, ensures that possible mistakes will be detected quickly and that the perpetrators will be located. The incident of the 41 Seveso drums does not contradict this claim. If the contractual relationship between consignor (waste generator) and consignee (waste disposer) is clear-cut and buttressed by verifiable documentation there is neither reason nor excuse for "misbehaviour" by the carrier. All those concerned with the development of new regulations should bear this axiom in mind as should those who must implement the subsequent rules arising from such regulations.

GENERAL STATEMENTS

1. Transfrontier movement of hazardous wastes must be justified. The prevalent attitude is that such wastes should be eliminated where they are generated. But if for reasons involving protection of the environment and/or if the installation of adequate facilities is not feasible or is incomplete in a certain region or country, then use of a facility in an other region may be appropriate, especially if required technology and capacity are available there. Only highly hazardous wastes require unusual methods and facilities (which cannot be provided in certain countries) for their elimination. Hence, the quantity of waste crossing frontiers may be limited. In the future, when improved technologies are employed, the direction of the waste streams may reverse. Even though environmental protection is the primary issue, expensive and unique facilities having the highest standards probably require a minimum input of material in order to be economically feasible. Therefore, such facilities might have to depend on foreign waste in order to operate efficiently. Thus, in Western Europe construction of more facilities than strictly needed for multinational regional pools would clearly be ill advised (special treatment plants, special incineration plants are included in this statement).

So far as the UTD is concerned, there is no other comparable underground waste disposal facility available in Europe at this time. Perhaps in the future, another mine somewhere may be converted into a waste disposal plant, provided environmental requirements can be met and the necessity for such capacity is clearly demonstrated (this logic also appears to be applicable in the case of treatment and storage of radioactive wastes). National borders should not be a barrier to countries who wish to develop the best possible solutions concerning treatment, disposal, and recycling operations involving hazardous wastes.

2.   Waste Generators and Waste Disposers are the Leading Actors.

Generators and disposers are responsible and liable in their field of endeavour today and for the future. This aspect has been discussed in great detail in other reports presented at this Seminar.

So far as the UTD is concerned the position is that UTD becomes liable upon acceptance of waste at the facility if the waste delivered is in accord with the specifications stated in Form A ("Formblatt A" of the UTD). The waste generator, however, is liable for any damage to the UTD resulting from the waste if the UTD can prove that the waste delivered was not in accord with the agreed specifications. The UTD can then demand re-importation of the waste by the generator at his expense.

3.   Before the waste generator requests the "confirmation" (as mentioned in step 3 under framework for export-import) from his governmental authority, technical procedures of investigation, identification and assessment concerning the waste with respect to the feasibility and desirability of underground disposal must be completed.

4.   Before the import licence application is submitted to the appropriate agency on behalf of the UTD, technical approval by the Mining Authority must be obtained and the contractual agreements between the waste generator and UTD must have been finalised.

Letter of acceptance: this document together with Form A, signed and validated by the Mining Authority is the basic documentation supporting the application for the import licence.

5.   All documents in this procedure must be prepared in German, the language of the importing country.

6.   UTD sees no necessity for any kind of "pre-notification", as proposed by several official entities. Such a requirement can only result in wasted time, money and energy. Adequate and timely information for all authorities concerned is available by means of the existing procedure.

7.   The "Import-Licence" (or permission) is an <u>indispensable requirement</u>. The "importing country" accepts the waste and with it any and all problems and administrative liabilities arising from its final whereabouts (storage, treatment, elimination, disposal) within the limits of its jurisdiction. The importing country thus must bear the burden of obtaining public acceptance of its activities. Before any real action can take place, public acceptance must be secured.

8.   Additional "papers" or more sophisticated manifests can only produce fictitious security. Such additional documents could actually reduce clarity of action and responsibility even though criminal intelligence will always find ways and means to defraud. But effective control of each step by those concerned, as provided and facilitated by clear requirements, can be achieved.

9.   Adequate environmental protection requires strict rules and ever improved regulations, especially insofar as hazardous waste is concerned. But the activities of waste generators and waste eliminators who provide disposal,

incineration, treatment and recycling ultimately decides the quality of the results. Therefore, at the outset, the waste generator must take care to cooperate directly with an appropriate and duly qualified disposal facility. This willingness to cooperate is the essential part of the generator's responsibility "from cradle to grave" for his hazardous waste. Third parties (broker, agent or others) cannot take the load off his shoulders, even if those parties are "licensed". Most waste generators have come to realise this situation and now proceed accordingly.

The waste generator decides the "what and why" in the course of his production as to how a waste may be generated. He may have to discontinue or suspend production if wastes are generated for which no environmentally safe elimination is feasible. And if the product is actually needed and cannot be dispensed with, the production processes may have to be changed. Such changes have occurred several times in the past and are increasingly likely to occur in the future.

CONCLUSIONS

Transfrontier movement of hazardous waste must be viewed as an exceptional case in the scope of national hazardous waste management policy. In connection with supra national environmental policy, certain reasons may exist (environmental, logistic and economic) to allow such movements. Proven tools are available to supervise and monitor such waste streams. The importing country must be in a position to approve or to reject foreign hazardous waste proposed for disposal in a facility within the limits of its jurisdiction. The importing country ultimately must justify potential impacts and implications which approval of the importation might create; public acceptance, effects on health and the environment, local disposal capacity and economic considerations must be fully taken into account. Close cooperation between the competent authorities of the exporting and importing countries promptly achieved at whatever level of detail is necessary may prevent irrational and unreasonable decisions. The economic interests of the waste generator and especially those of the waste disposer can not be the decisive factors in reaching such decisions.

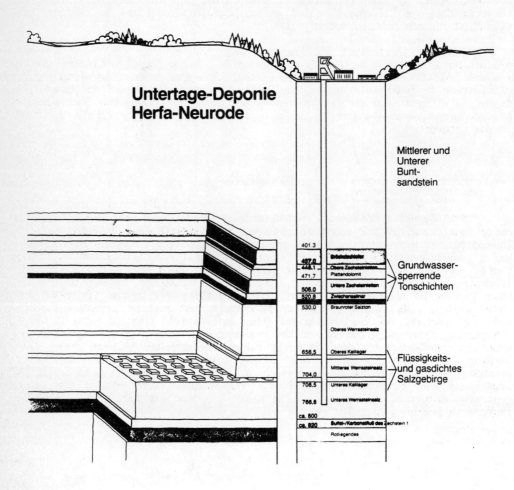

Untertage-Deponie Herfa-Neurode

Mittlerer und Unterer Bunt-sandstein

Grundwasser-sperrende Tonschichten

Flüssigkeits-und gasdichtes Salzgebirge

Tiefe	Schicht
401.3	
437.0	Bröckelschiefer
448.1	Oberes Zechsteinletten
471.7	Plattendolomit
505.0	Untere Zechsteinletten
520.8	Zwischenanhydrit
530.0	Braunroter Salzton
	Oberes Werrasteinsalz
656.5	Oberes Kalilager
	Mittleres Werrasteinsalz
704.0	
708.5	Unteres Kalilager
766.8	Unteres Werrasteinsalz
ca. 800	
ca. 820	Sulfat-/Karbonatfuß des Zechstein 1
	Rotliegendes

Figure 1

168

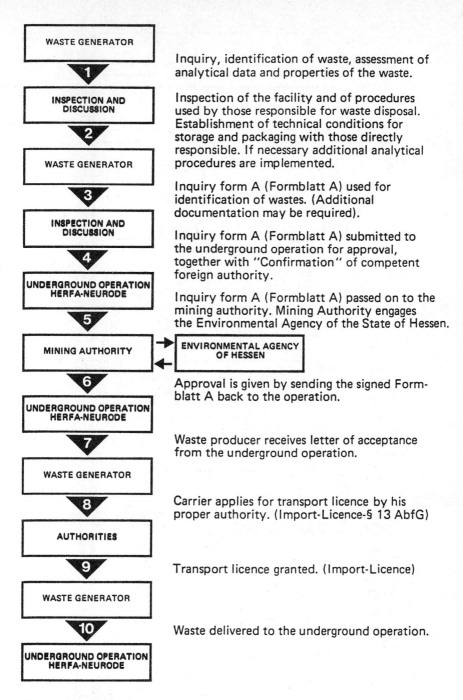

WASTE GENERATOR	Inquiry, identification of waste, assessment of analytical data and properties of the waste.
1	
INSPECTION AND DISCUSSION	Inspection of the facility and of procedures used by those responsible for waste disposal. Establishment of technical conditions for storage and packaging with those directly responsible. If necessary additional analytical procedures are implemented.
2	
WASTE GENERATOR	
3	Inquiry form A (Formblatt A) used for identification of wastes. (Additional documentation may be required).
INSPECTION AND DISCUSSION	Inquiry form A (Formblatt A) submitted to the underground operation for approval, together with "Confirmation" of competent foreign authority.
4	
UNDERGROUND OPERATION HERFA-NEURODE	Inquiry form A (Formblatt A) passed on to the mining authority. Mining Authority engages the Environmental Agency of the State of Hessen.
5	
MINING AUTHORITY → ← **ENVIRONMENTAL AGENCY OF HESSEN**	
6	Approval is given by sending the signed Formblatt A back to the operation.
UNDERGROUND OPERATION HERFA-NEURODE	
7	Waste producer receives letter of acceptance from the underground operation.
WASTE GENERATOR	
8	Carrier applies for transport licence by his proper authority. (Import-Licence-§ 13 AbfG)
AUTHORITIES	
9	Transport licence granted. (Import-Licence)
WASTE GENERATOR	
10	Waste delivered to the underground operation.
UNDERGROUND OPERATION HERFA-NEURODE	

Fig. 2 — Procedure for acceptance and licence
for each particular waste

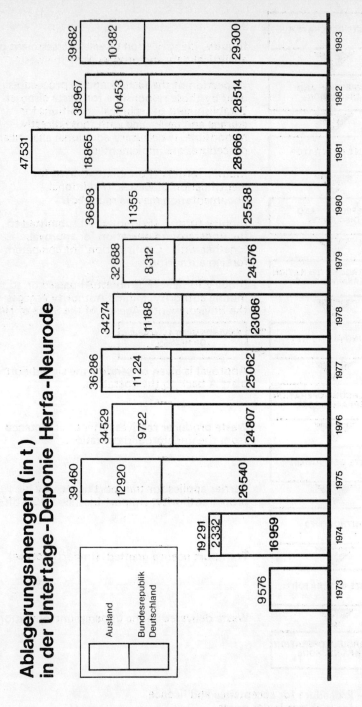

Ablagerungsmengen (in t) in der Untertage-Deponie Herfa-Neurode

Legend:
- Ausland
- Bundesrepublik Deutschland

1983: 39 682 — 10 382 / 29 300
1982: 38 967 — 10 453 / 28 514
1981: 47 531 — 18 865 / 28 666
1980: 36 893 — 11 355 / 25 538
1979: 32 888 — 8 312 / 24 576
1978: 34 274 — 11 188 / 23 086
1977: 36 286 — 11 224 / 25 062
1976: 34 529 — 9 722 / 24 807
1975: 39 460 — 12 920 / 26 540
1974: 19 291 — 2 332 / 16 959
1973: 9 576

Figure 3

<u>A N N E X</u>

<u>Samples of UTD documents</u>

A.1.1

Untertage-Deponie Herfa-Neurode

Sehr geehrte Damen und Herren!

Wir übersenden Ihnen die „**Allgemeinen Geschäftsbedingungen**" und das „**Formblatt A**" unserer Untertage-Deponie Herfa-Neurode.

In der Untertage-Deponie Herfa-Neurode werden unvermeidbare Abfälle aus Industrie und Gewerbe abgelagert, deren sichere, umweltunschädliche Beseitigung mit anderen Beseitigungsverfahren nicht möglich oder problematisch ist. Dies gilt insbesondere für einen Teil der Abfälle, die im Sinne des § 2 (2) des Abfallbeseitigungsgesetzes bestimmt sind und für solche Abfälle, bei denen durch wasserlösliche oder andere Anteile eine potentielle Gefahr für die Umwelt nicht ausgeschlossen werden kann.

Die notwendige Gewährleistung der Sicherheit des Untertage-Betriebes erfordert jedoch die folgenden Einschränkungen bei der Annahme für die Ablagerung:

1. Die Abfälle dürfen unter den Ablagerungsbedingungen in den praktisch geschlossenen Räumen unter Tage keine explosiven, zündfähigen oder gefährlichen Gas-Luft-Gemische bilden.

2. Abfälle, die zur Selbstentzündung neigen, können nicht abgelagert werden.

3. Flüssige Abfälle können ebenfalls nicht abgelagert werden. Der Beseitigungspflichtige muß gegebenenfalls bei solchen Abfällen eine absolut stichfeste Konsistenz durch geeignete Behandlungsmethoden herstellen. Bei Beschädigung der Verpackung dürfen keine freien Flüssigkeiten austreten können.

4. Für die sichere Handhabung der Abfälle bei Transport und Ablagerung ist eine einwandfreie Verpackung und Palettierung auf Einweg-Paletten erforderlich. Mechanisch intakte, äußerlich saubere, gebrauchte Stahlblechfässer mit Spannringverschluß haben sich bewährt. Andere Behälter oder Verpackungen können — abhängig von den Eigenschaften des Abfalls — von Fall zu Fall vereinbart werden (Kunststoff-Fässer u. a.).

Bei vielen Abfällen, die zunächst unter diese Beschränkungen fallen und deren anderweitige umweltunschädliche Beseitigung technisch und wirtschaftlich nicht gegeben ist, können oft durch gemeinsam zu vereinbarende Verfahren (z. B. Konditionierung, besondere Verpackung, Abmauerung unter Tage, Einbetonieren usw.) die Voraussetzungen für die sichere Ablagerung in der Untertage-Deponie geschaffen werden.

Mit der Übersendung des "Formblatts A" (in zweifacher Ausfertigung) an die Kali und Salz AG, Abt. BOD, Postfach 10 20 29, 3500 Kassel, erkennt der Beseitigungspflichtige die „Allgemeinen Geschäftsbedingungen" an.

b. w.

Genehmigungsverfahren

Der Beseitigungspflichtige (Abfallerzeuger bzw. Besitzer des Abfalls) erhält die **schriftliche Annahme-bestätigung** nach Prüfung des betreffenden Abfalls und der Zustimmung durch das Bergamt Bad Hers-feld. Eine Kopie des mit dem Zustimmungsvermerk des Bergamtes versehenen „Formblatts A" wird beigefügt.

In der **Annahmebestätigung** und auf dem „**Formblatt A**" wird von der Untertage-Deponie die Code-Bezeichnung angegeben, mit der die Behälter, in denen der betreffende Abfall angeliefert wird, deut-lich lesbar und dauerhaft zu beschriften sind.

Annahmebestätigung, bergbehördliche Zustimmung und Code-Bezeichnung gelten für die wiederholte Anlieferung des gleichen Abfalls.

Mit diesen Unterlagen kann der Beseitigungspflichtige oder das von diesem beauftragte Transport-unternehmen die Transport- bzw. Importgenehmigung bei der zuständigen Behörde beantragen. (S. Verordnungen der Bundesregierung zu den §§ 11, 12 und 13 des Abfallbeseitigungsgesetzes)

Nach Erhalt der Genehmigungen vereinbart der Beseitigungspflichtige bzw. das beauftragte Transport-unternehmen die verbindlichen Anlieferungstermine mit dem Betrieb der Untertage-Deponie Herfa-Neurode.

In der **schriftlichen Annahmebestätigung** der Untertage-Deponie Herfa-Neurode sind die weiteren Einzelheiten der Anlieferung und die gegebenenfalls vereinbarten „Besonderen Bedingungen" für Anlieferung, Verpackung und Ablagerung festgelegt.

Für die sichere Beseitigung von problematischen Abfällen ist eine enge Zusammenarbeit zwischen dem Beseitigungspflichtigen und dem Beseitiger unerläßlich. Dies erfordert die notwendige und hinreichende gegenseitige Kenntnis der Gegebenheiten bei der Entstehung und Behandlung des Abfalls sowie die der Voraussetzungen und Bedingungen für die sichere Ablagerung in der Untertage-Deponie.

Wir stehen Ihnen für weitere Auskünfte zur Verfügung, falls durch die Ablagerung Ihrer Abfälle in der Untertage-Deponie eine abfallwirtschaftlich sinnvolle Lösung erreicht werden kann.

Mit freundlichem Gruß und Glückauf
KALI UND SALZ AKTIENGESELLSCHAFT

Annahmebestätigung

In der Anlage übersenden wir die mit dem Vermerk des Bergamtes Bad Hersfeld versehene Fotokopie des Formblatts A für den Abfall mit der

Code-Nr...............................

Wir erklären uns bereit, diesen Abfall gemäß den beigefügten „**Allgemeinen Geschäftsbedingungen**" abzulagern.

Diese Annahmebestätigung gilt für die wiederholte Anlieferung dieses Abfalls. Den Antrag zur Genehmigung nach den Verordnungen zu § 12 (Transport) oder § 13 (Import) des AbfG hat der Beseitigungspflichtige bzw. das von ihm beauftragte Transportunternehmen bei der zuständigen Behörde unter Beifügung einer Kopie dieser Unterlagen zu stellen.

Jeder Behälter ist deutlich und dauerhaft **mit der Code-Nr. zu beschriften.** Alle anderen Beschriftungen der Behälter, die mit dem Inhalt nicht übereinstimmen, müssen entfernt werden.
Bei Behältern, die nicht mechanisch einwandfrei, äußerlich sauber und verschlossen sind und die den Angaben des Formblatts A und den unter „Besondere Bedingungen" festgelegten Vereinbarungen nicht entsprechen, kann die Annahme verweigert werden.

Die Anlieferung muß auf Einweg-Paletten erfolgen (max. 1200 x 1200 mm, max. Höhe 1100 mm). Auf einer Palette dürfen nur Behälter einer Größe und gleicher Code-Nr. verladen werden. Die Behälter sind auf den Paletten durch Stahlbänder zuverlässig gegen Verrutschen zu sichern.

Auf den Begleitscheinen müssen zusätzlich **Code-Nr., Faßzahl** sowie bei gleichzeitiger Anlieferung mehrerer Abfallarten die **Gewichte für die jeweilige Code-Nr.** angegeben sein.

Bei Anlieferungen unter 1000 kg wird der Preis für eine Tonne berechnet.

Die Anlieferungstermine sind telefonisch mit uns **zu vereinbaren** (Tel. Durchwahl 06624/81308, nachmittags ab 15.00 Uhr 81325). Die Anlieferung kann nur in der Zeit von 5.30 Uhr bis 13.00 Uhr erfolgen. Eine Annahme nach 13.00 Uhr ist nur in Ausnahmefällen nach vorheriger Einholung unseres Einverständnisses (Tel. 06624/81308) möglich.

Besondere Bedingungen:

UNTERTAGE-DEPONIE HERFA-NEURODE
der
Kali und Salz AG

3500 Kassel, Friedrich-Ebert-Straße 160
Postfach 102029
Telefon: (0561) 30 11 (Durchwahl 301–395)
Telex: 992418, 992392

6432 Heringen (Werra), Werk Wintershall
Betrieb; Untertage-Deponie Herfa-Neurode
Telefon: (06624) 81–1
Telex: 4 93 383

Dieses **Formblatt A** ist vom Beseitigungspflichtigen in doppelter Ausfertigung an Kali und Salz AG, Abt. BOD, Postfach 102029, 3500 Kassel, zu übersenden. Die schriftliche Annahmebestätigung erfolgt nach Überprüfung der Angaben.

1. **Bezeichnung des Abfalles, Menge, Art der Verpackung** (ggf. Stoffkennzeichnung nach ADR bzw. EVO):

2. **Produktionsherkunft** (Verfahren)
2. a) Ortsangabe (Land, PLZ) des Betriebes, in dem der Abfall entstanden ist.
3. **Codebezeichnung** (Mit K + S zu vereinbaren):

4. a) Wie wurde der Abfall nach seinem Anfallen beim Beseitigungspflichtigen behandelt und gelagert?

b) War oder ist der Abfall Gegenstand eines Bußgeld- bzw. Strafverfahrens oder öffentlicher Diskussion: ja / nein
(Wenn ja, bitte Angaben dazu auf gesondertem Blatt beifügen.)

5. Chemische Bezeichnung der Einzelkomponenten	Ø	min.	Anteile in % max.
1.			
2.			
3.			
4.			
5.			
6.			
7.			
8.			
9.			
10.			

6. **Beschaffenheit (bei 30°C):**

flüssig ☐	fest erstarrte Schmelze ☐	schlammig/stichfest ☐
zähflüssig/teigig ☐	fest-körnig ☐	schlammig/flüssig ☐
	fest-pulverig ☐	

7. Wasser- bzw. Flüssigkeitsgehalt (Gewichts-%) _____

Aus welchen Stoffen bestehen die Flüssigkeiten?

8. Schmelzpunkt (°C) _____   11. PH-Wert   _____

9. Siedepunkt   (°C) _____   12. Gasdruck bei 20°C (Torr) _____

10. Flammpunkt   (°C) _____   bei 50°C (Torr) _____

13. **Welche Gase** kann der Abfall durch eventuelle Nachreaktionen unter den Ablagerungsbedingungen entwickeln?

a) Wenn er im Anlieferungsbehälter eingeschlossen bleibt:

b) Wenn er mit Luft in Berührung kommt:

c) Wenn er mit dem anstehenden Salz in Berührung kommt:

d) Bei welchen Temperaturen treten Zersetzungen, Ausgasungen, spontane Zersetzungen oder Explosionen auf?

Welche Gasgemische können bei den betreffenden Temperaturen entstehen?

A.3.2

14. **Angaben über die toxikologischen Eigenschaften des Abfalles** (ggf. ausführliche Angaben gesondert beifügen)

15. **Bei Transportschäden,** insbesondere Bränden:

a) Geeignete Löschmittel, unzulässige Löschmittel

b) Atemschutz

c) Angaben über Vorschriften (behördliche und/oder werksinterne) zur Behandlung von Personen, die mit dem Abfall direkt in Berührung gekommen sind (Schleimhäute, Haut) oder die bei Bränden den entstehenden Gasen ausgesetzt waren (ggf. ausführliche Angaben gesondert beifügen)

Angaben zu den Bedingungen unter Tage der Deponie Herfa-Neurode

Die Abfälle werden in leergeförderten Abbauen von etwa 2,5–3,5 m Höhe abgelagert. Temperatur: 25–30° C, relative Luftfeuchtigkeit: max. 45%. Gehalte der Grubenwetter an Abgasen: $CO: \sim 0{,}001$ (Vol.%), $CO_2: \sim 0{,}1$ (Vol. %), $NO_2: \sim 2{,}5$ (mg/m³).
Das anstehende Salz hat im Mittel folgende Zusammensetzung:

$MgSO_4 \cdot H_2O$ (Kieserit): 10–20%	$KCl \cdot MgCl_2 \cdot 6H_2O$ (Carnallit): 5–10%	Unlösliches (Ton usw.): 1–1,5%
KCl (Sylvin): 10–15%	NaCl (Steinsalz): 60–65%	

Erklärung

Wir versichern, daß die im Formblatt A zum Vertrag mit der K+S gemachten Angaben zutreffen.
Die zur Ablagerung in die Untertage-Deponie Herfa-Neurode anzuliefernden Abfälle entsprechen den aufgeführten Deklarationen. Das mit der Deklaration und dem Transport beauftragte Personal ist von uns gegen Unterschriftsbestätigung darauf hingewiesen worden, daß
a) nur genau definierte Abfälle entsprechend den Angaben dieses Formblattes A zur Abfuhr bereitgestellt,
b) keine anderen als die im Formblatt A definierten Abfälle zur Untertage-Deponie Herfa-Neurode angeliefert werden dürfen.

Verantwortlich	Name	Telefon	Name des Stellvertreters
a) Für die analytischen Angaben:			
b) Für gewissenhafte Deklaration:			
c) Für Verladung und Transport:			

Anschrift   Datum: Rechtsverbindliche Unterschrift
des Beseitigungspflichtigen: des Beseitigungspflichtigen

Raum für Behördenvermerke:

UNTERTAGE-DEPONIE HERFA-NEURODE

der

Kali und Salz AG

Allgemeine Geschäftsbedingungen (1. 1. 1983)

§ 1

Die Kali und Salz Aktiengesellschaft — im folgenden „K + S" genannt — übernimmt die in Formblatt A beschriebenen Abfälle des Beseitigungspflichtigen — im folgenden „Firma" genannt — zur Ablagerung in ihrer Untertage-Deponie nach Maßgabe ihrer schriftlichen Annahmebestätigung sowie dieser Geschäftsbedingungen.

§ 6

(1) Vor Beginn der Abladung muß die Genehmigung nach § 12 bzw. § 13 des Abfallbeseitigungsgesetzes nachgewiesen werden.

§ 9

(1) Unbeschadet der gesetzlichen Bestimmungen haftet die Firma für jeden Schaden, der K + S dadurch entsteht, daß

a) das Formblatt A unrichtig oder unvollständig ausgefüllt ist,

b) der Abfall und dessen Eigenschaften nicht den Angaben des Formblatts A entspricht,

c) der Abfall nicht gemäß Formblatt A oder den mit K + S in der schriftlichen Annahmebestätigung getroffenen Vereinbarungen verpackt ist.

(2) Die Firma stellt K + S von allen Schadenersatzansprüchen Dritter frei, die auf den in Abs. (1) a — c genannten Gründen beruhen.

§ 10

(1) K + S haftet für schuldhaftes Handeln, jedoch begrenzt auf die Höhe der Deckungssumme der von ihr abgeschlossenen Haftpflichtversicherung.

(2) In demselben Umfang stellt K + S die Firma von allen Ansprüchen Dritter frei, soweit sie auf schuldhaftem Handeln von K + S (oder ihren Erfüllungs- oder Verrichtungsgehilfen) bei der Erfüllung dieses Vertrages beruhen.

§ 11

(1) K + S ist über die gesetzlichen Bestimmungen und über den Fall des § 6 dieses Vertrages hinaus von ihrer Ablagerungsverpflichtung gemäß § 1 befreit, wenn

a) die Firma für die jeweilige Sendung nach § 2 (2) keinen Anlieferungstermin vereinbart hat,

b) einer der in § 9 Abs. (1) a — c genannten Gründe vorliegt,

c) die Ablagerung durch Gesetz oder durch behördliche oder gerichtliche Anordnung untersagt wird,

d) die Ablagerung für K + S aus betrieblichen oder sonstigen wichtigen Gründen, die durch den Vorrang des Kaliabbaus im Werk Wintershall (Mineralgewinnung) bedingt sind, unmöglich oder unzumutbar erschwert ist,

e) durch die beabsichtigte oder vorgenommene Ablagerung berechtigte Interessen von K + S erheblich beeinträchtigt werden (z. B. durch Widerstand in der Öffentlichkeit), es sei denn, daß ein Verschulden von K + S vorliegt.

(2) Die Befreiung von der Ablagerungsverpflichtung gilt nur für die jeweiligen Teilmengen des abzulagernden Abfalls, für die einer der in Abs. (1) genannten Gründe eingetreten ist.

§ 12

(1) Die Pflicht von K + S zur Verwahrung des abgelagerten Abfalls besteht so lange, bis der Abfall in das Eigentum von K + S übergeht.

(2) K + S kann, unbeschadet der gesetzlichen Bestimmungen, die Rücknahme abgelagerter Abfälle nur dann verlangen, wenn einer der in § 9 Abs. (1) a — c genannten Gründe oder ein anderer wichtiger Grund vorliegt, der nicht in die Risikosphäre von K + S fällt. Mit Eigentumsübergang erlischt die vertragliche Rücknahmeverpflichtung.

(3) Übt K + S einen Rücknahmeanspruch gemäß Abs. (2) aus einem Grunde aus, den die Firma zu vertreten hat, so hat die Firma alle sich daraus ergebenden Kosten zu tragen.

(4) Ein Anspruch auf Erstattung der gezahlten Vergütungen im Falle der Ausübung des Rücknahmeanspruchs gemäß Abs. (2) besteht nicht.

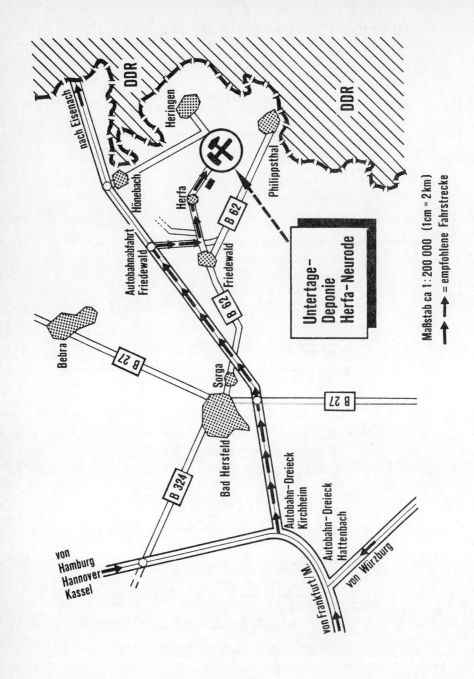

DDR

DDR

nach Eisenach

Heringen

Hönebach

Herfa

B 62

Philippsthal

Autobahnabfahrt
Friedewald

B 62 Friedewald

Bebra

B 27

Sorga

B 27

Bad Hersfeld

B 324

Autobahn–Dreieck
Kirchheim

Autobahn–Dreieck
Hattenbach

von
Hamburg
Hannover
Kassel

von Frankfurt / M.

von Würzburg

Untertage–
Deponie
Herfa–Neurode

Maßstab ca 1 : 200 000 (1cm = 2km)

= empfohlene Fahrstrecke

A.6.1

Antrag auf Erteilung einer Genehmigung zur Einfuhr von Abfällen nach § 13 Abfallbeseitigungsgesetz in Verbindung mit § 1 der Abfalleinfuhr-Verordnung

Zutreffendes ist angekreuzt: ☒

1. **Antragsteller**
Name, Firma:

_____

Anschrift: (Postleitzahl, Wohnort, Straße und Hausnummer)

_____

| Staat | Telefon | Registr.-Nr. (soweit amtlich festgelegt) |

_____

Frühere Anträge

Sind früher (auch von anderen Behörden im Geltungsbereich dieser Verordnung) bereits Genehmigungen nach § 13 Abfallbeseitigungsgesetz erteilt oder versagt worden? ja ○ nein ○
Falls ja, nähere Angaben (ggf. besonderes Blatt)

_____

2. **Die Genehmigung soll gelten**

– für eine einzelne Verbringung ○; Beginn: _____

– ohne Beschränkung auf eine einzelne Verbringung im Zeitraum von _____ bis _____ ○

3. **Abfallbeschreibung**

3.1 **Abfallerzeuger**
Name, Firma:

_____

Anschrift: (Postleitzahl, Wohnort, Straße und Hausnummer)

_____

| Staat | Telefon |

_____

| Werk/Betriebsteil | Registr.-Nr. (soweit amtlich festgelegt) |

_____

3.2 **Bezeichnung der Abfälle**
Übliche Bezeichnung

_____

wissenschaftl.-techn. Bezeichnung (ggf. Erläuterung auf besond. Blatt)

_____

| Nr. des Abfallschlüssels | Bundesland, das Abfallschlüssel amtlich eingeführt hat (soweit zutreffend) |

_____

3.3 **Technische Herkunft der Abfälle**
Im Produktionsbereich

_____

Angefallen bei Produktion von

_____

Verwendete Roh- und Grundstoffe

_____

Für die Abfallzusammensetzung wesentliche Hilfsstoffe

_____

3.4 **Beschaffenheit der Abfälle**
Erscheinungsbild: fest ○

staubförmig ○   granuliert ○   brockig ○   sperrig ○

pastös ○   breiig ○ _____% Feststoffgehalt

flüssig ○

| Farbe | Geruch |

_____

7U/5U6 – Deutscher Gemeindeverlag GmbH – 10/74
704 20/04 – Formularverlag W. Kohlhammer

A.6.2

Besondere Eigenschaften: giftig ○ krankheitserregend ○ ätzend ○ feuergefährlich ○ selbstentzündlich ○
stark riechend ○. ekelerregend ○

Kurze chemisch-physikalische Kennzeichnung: organisch: sauer ○ basisch ○
anorganisch: sauer ○ basisch ○

cyanhaltig ○ metallsalzhaltig ○ halogenhaltig ○ leichtflüchtig ○ brennbar ○

Aufzählung aller Bestandteile und Inhaltsstoffe

_____

_____

Welcher Stoff bildet den flüssigen Anteil?

_____%

Wie groß ist der wasserlösliche Anteil?_____%

leichtlöslich ○ schwerlöslich ○

Ist der wasserlösliche Anteil giftig? ja ○ nein ○

Beschreibung aller Gefahren, die von den Abfällen ausgehen:

_____

_____

Mögliche Reaktionen mit anderen Stoffen:

_____

_____

Sonstige Hinweise auf die Beschaffenheit der Abfälle, eigene Erfahrungen usw.

_____

_____

3.5 Chemisch-physikalisch-biologische Analyse (§ 1 Abs. 3 der Verordnung)

Erstellt durch: Name des Instituts/Sachverständigen

_____

Anschrift: (Postleitzahl, Wohnort, Straße und Hausnummer)

_____

Staat   Telefon

_____

3.6 Schutzmaßnahmen

Sind Schutzmaßnahmen erforderlich? ja ○ nein ○

Falls ja, genaue Beschreibung:

_____

Sind die Abfälle gefährliches Gut im Sinne der einschlägigen nationalen oder internationalen Vorschriften über die Beförderung gefährlicher Güter? ja ○ nein ○

Vorschrift   Klasse   Ziffer

3.7 Anlieferung: unverpackt ○ in Gebinden aus Holz ○ Metall ○ Kunststoff ○ Papier ○
in Säcken aus Kunststoff ○ Gewebe ○ Papier ○
in Fässern offen ○ geschlossen ○ in Mulden offen ○ geschlossen ○
in Containern offen ○ geschlossen ○ in Tankwagen ○

in sonstiger Weise (bitte beschreiben) ○

4. Menge der Abfälle

– bei der einzelnen Verbringung _____ t _____ m³

– bei mehrfachen Verbringungen je Verbringung _____ t _____ m³

insgesamt _____ t _____ m³

5. Beförderung

5.1 Beförderer

Name, Firma:

_____

Anschrift: (Postleitzahl, Wohnort, Straße und Hausnummer)

_____

Staat   Telefon   Registr.-Nr. (soweit amtlich festgelegt)

_____

180

A.6.3

5.2 Frühere Abfalltransporte des Beförderers

Sind früher (auch von anderen Behörden im Geltungsbereich dieser Verordnung) bereits Genehmigungen nach §§ 12 oder 13 Abfallbeseitigungsgesetz erteilt oder versagt worden.   ja ○   nein ○
Falls ja, nähere Angaben (ggf. besond. Blatt)

_____

_____

5.3 Beförderungsablauf

Übernahme der Abfälle am _____ Ort: _____

Grenzübergang am   _____ Übergangsstelle: _____
Beförderungsweg innerhalb des Geltungsbereichs des Abfallbeseitigungsgesetzes (mit Zeitangaben):

_____

Beförderungsmittel (mit besonderem Hinweis auf vorgesehene Umladungen):

_____

Übergabe der Abfälle am   _____ Ort: _____
an: (Name und Anschrift: Postleitzahl, Wohnort, Straße und Hausnummer)

6. Abfallbeseitigungsanlagen
Soll die Beseitigung der Abfälle durch mehrere Abfallbeseitigungsanlagen neben- oder hintereinander erfolgen, sind die nach-
folgenden Angaben für jede der beteiligten Anlagen und ihre Betreiber zu machen. Aus der Beantwortung der Nr. 6.3 muß sich
der Ablauf der gesamten Beseitigung nach Art und Reihenfolge eindeutig ergeben (ggf. besond. Blatt)

6.1 Betreiber
Name, Firma:

_____

Anschrift: (Postleitzahl, Wohnort, Straße und Hausnummer)   Telefon

_____

Name, Firma:

_____

Anschrift: (Postleitzahl, Wohnort, Straße und Hausnummer)   Telefon

_____

6.2 Anlage

| Standort | Telefon | Registr.-Nr. (soweit amtlich festgelegt) |

| Standort | Telefon | Registr.-Nr. (soweit amtlich festgelegt) |

6.3 Vorgesehene Art der Beseitigung

6.3.1 Lagerung ○; Dauer: _____
Angaben, wann und wie die Abfälle später behandelt oder abgelagert werden sollen, sind in Nr. 6.3.2 und 6.3.3 zu machen.

6.3.2 Behandlung: Verbrennung ○   Kompostierung ○   Sonstiges (nähere Angaben hierzu erforderlich) ○

_____

Beginn: _____

6.3.3 Ablagerung ○
Zusätzliche Erläuterungen zum Ablauf der Beseitigung

_____

Ort, Datum   Unterschrift

_____

Erklärung des Betreibers der Abfallbeseitigungsanlage
Ich erkläre, die Richtigkeit der Angaben des Antragstellers zu meiner Person und der von mir betriebenen Abfallbeseitigungsan-
lage. Ich bin bereit, die in diesem Antrag nach Art und Menge beschriebenen Abfälle zu übernehmen. Ich versichere die Eignung
meiner Anlage zur schadlosen und geordneten Beseitigung der Abfälle in der im Antrag beschriebenen Weise.

Ort, Datum   Unterschrift und Stempel des Betreibers der Anlage

_____

_____

A.6.4

(Stempel der zuständigen Behörde)

Genehmigungsvermerk

1. Der vorstehende Antrag wird unter folgenden Bedingungen und Auflagen genehmigt (Zutreffendes ist angekreuzt):

1.1 Der Genehmigungsvermerk oder ein beglaubigter Abdruck ist in allen zum Verbringen der Abfälle benutzten Fahrzeugen mitzuführen.   ◯

1.2 Abfälle, die nach Arten getrennt gelagert, abgelagert oder behandelt werden sollen, sind getrennt einzusammeln und zu befördern.   ◯

1.3 Das Bedienungspersonal ist in geeigneter Weise mit den bei der Verbringung von Abfällen zusammenhängenden Gefahren vertraut zu machen.   ◯

1.4 Wesentliche Änderungen der im Antrag gemachten Angaben sind der genehmigenden Dienststelle unverzüglich mitzuteilen.   ◯

1.5 Ggf. weitere Bedingungen und Auflagen (z.B. Beförderungsweg):

2. **Die im Antrag gemachten Angaben sind Bestandteil dieser Genehmigung.**

3. **Hinweise**

3.1 Bei der Verbringung der Abfälle sind alle einschlägigen Vorschriften, insbesondere der Grundsatz des § 2 des Abfallbeseitigungsgesetzes, zu beachten.

3.2 Die beförderten Abfälle dürfen nur bei der im Antrag genannten Anlage angeliefert werden.

3.3 Auflagen und Bedingungen können nachträglich geändert oder ergänzt werden.

3.4 Diese Genehmigung kann bei unrichtigen oder unvollständigen Angaben im Antrag, bei Nichteinhaltung der Auflagen oder bei sonstigen Verstößen gegen die Vorschriften des Abfallbeseitigungsgesetzes und die dazu ergangenen Durchführungsbestimmungen aufgehoben werden. Außerdem können Verstöße gegen diese Vorschriften als Straftaten oder Ordnungswidrigkeiten (§§ 16, 18 des Abfallbeseitigungsgesetzes) geahndet werden.

3.5 Ggf. weitere Hinweise:

4. **Dieser Bescheid ist gebührenpflichtig. Die Gebühr wird gemäß § 5 der Abfalleinfuhr-Verordnung auf   DM festgesetzt.**

Rechtsbehelfsbelehrung:
Gegen diesen Bescheid kann innerhalb eines Monats nach Bekanntgabe Widerspruch erhoben werden. Der Widerspruch ist schriftlich oder zur Niederschrift einzulegen bei: (Behörde, Ort, Straße, Nr.)

– Falls die Frist durch das Verschulden eines von Ihnen Bevollmächtigten versäumt werden sollte, so würde dessen Verschulden Ihnen zugerechnet werden. –

Ort, Datum, Unterschrift der Behörde

A.7

MINISTERE DE L'INDUSTRIE ET DE LA RECHERCHE

DIRECTION INTERDEPARTEMENTALE DE L'INDUSTRIE

A T T E S T A T I O N

Nous confirmons que le déchet d'anneaux Raschig imprégnés de boues arsénieuses provenant de tours Vetrocoke d'absorption de CO_2, mentionné dans le formulaire A de la société Kali und Salz, numéro de code CFC-004, a été correctement et intégralement déclaré.

Nous recommandons le stockage souterrain de ce déchet dans la décharge de Herfa-Neurode.

Il n'existe en France actuellement, à ma connaissance, aucune possibilité équivalente de procéder à une élimination sûre et non polluante de ce déchet.

La possibilité d'utiliser les composants de ces déchets a été envisagée. Une réutilisation même partielle n'est pas actuellement possible.

P/LE DIRECTEUR, et par délégation
L'INGENIEUR DES MINES, Chef de la
DIVISION ENVIRONNEMENT

B E S T Ä T I G U N G

Wir bestätigen, daß der Abfall "mit Arsen oxyd verunreinigte Raschigringe aus dem Vetrocoke-Prozeß" im Formblatt A der Kali und Salz A.G., Code-Nummer CFC-004 richtig und vollständig deklariert ist.

Wir befürworten, daß dieser Abfall in der Untertage-Deponie Herfa-Neurode abgelagert wird.

Zur Zeit gibt es für diesen Abfall in Frankreich kein gleichwertige Möglichkeit der sicheren und umweltunschädliche Beseitigung.

Aufgrund der im Abfall enthaltenen stoffkomponenten wurde die Moeglichkeit einer Verwertung geprueft. Eine verwertung - auch von teilkomponent ist zur zeit nicht moeglich.

P/LE DIRECTEUR, et par délégation
L'INGENIEUR DES MINES, Chef de la
Division Environnement

183

Begleitschein Beleg zum Nachweis der Beseitigung von Abfällen

Dieser Beleg (weiß) ist mit Unterschrift des Beförderers im Nachweisbuch des Abfallerzeugers abzuheften.

Nr.: 1000 526871 ₁₀ | 1 |₁₁

30 Paletten ① Abfallart 120 Fässer

Filterationsrückstände BAB-002

② Abfallschlüssel-nummer ₁₆|₁₇

③ Abfallmenge m³ ₂₂|₂₃ t ₂₈

④ Konsistenz: fest = 1 stichfest = 2 pastös/schlammig/breiig = 3 staubförmig = 4 flüssig = 5

⑤ Amtl. Kennzeichen des Fahrzeuges: HEF – ZL

⑥ Art des Fahrzeugs: Lkw/Container = 1 Tankfahrzeug = 2 Bahn/Kesselwagen = 3 sonstiges Fahrzeug = 4

⑦ Betriebsnummer ₃₈ | S 1 3 0 3 8 3 |₃₁

⑧ Beförderernummer ₄₆ | F 7 3 T 0 0 0 1 |₃₉

⑨ Beseitigernummer ₅₄ | F 7 3 B 1 0 0 1 |₄₇

⑩ Abfallerzeuger (Name, Anschrift oder Stempel)
BASF
Antwerpen N.V.
2040 Antwerpen

⑪ Abfallbeförderer (Name, Anschrift oder Stempel)
Krug
Internationale Spedition
& Handelsgesellschaft mbH
6445 Alheim-Heinebach
Tel.: ...
Datum der Übernahme

⑫ Abfallbeseitiger (Name, Anschrift oder Stempel)
Kali & Salz AG
UTD Herfa/Neurode
Heringen/Werra

⑬ Versicherung der richtigen Deklarierung
Datum der Ausstellung Tag ₅₅ Monat Jahr 60

⑭ Versicherung der ordnungs-gemäßen Beförderung
Tag 61 Monat Jahr 66

⑮ Versicherung der Annahme zur ordnungsgemäßen Beseitigung
Datum der Annahme Tag 67 Monat Jahr 72

Unterschrift Unterschrift Unterschrift
Frei für betriebsinterne Vermerke

WEKA-VERLAG, Industriestraße 21, D-8901 Kissing, Telefon 0 82 33 / 2 30, Telex 533 287 weka d. Bestell-Nr.: 4402 – Begleitschein gem. § 2 Abs. 1 AbfNachwV, Durchschreibesatz à 6 Blatt, Blatt 1

FINANCIAL SECURITY AS A MEANS TO IMPROVE CONTROL OF TRANSFRONTIER MOVEMENTS OF HAZARDOUS WASTE

Henri Smets
Environment Directorate
OECD, Paris

SUMMARY

As a back-up to control of transfrontier movements of hazardous waste, a system of financial security, to be paid out automatically by the carrier in the event of loss of waste shiments, would have a deterrent effect and would very largely preclude illegal practices that may harm the environment. The system would entail no extra cost and no extra administrative expenditure provided the waste reached its destination. Instituted under an international agreement, the system would be equivalent in effect to uniform strengthening of penalties and unification of rules on liability, without in fact amending either. Indirectly the system could reduce administrative controls and would ultimately exclude unreliable operators from the market.

INTRODUCTION

A particular feature of the transport of hazardous waste from the generator to the disposer is that as a rule the value of the shipment is negative: if the shipment were to disappear, transport costs would be cut and disposal costs would not arise. These costs total around $100 per tonne, so there is some incentive to mislay a shipment when the risk (to the parties concerned) is not very high.

Unethical operators have already been convicted of tipping hazardous waste instead of disposing of it properly. But it is very difficult to supervise every loading, storage and unloading operation, to identify those responsible for illegal tipping, and to compel them to pay for cleaning up and disposal when the waste is finally located. It may accordingly be advantageous to consider a system that inhibits illegal operations with their damaging consequences.

This paper examines how financial security might help to improve the control of transfrontier movements of hazardous waste. It first describes the proposed system, and then considers its advantages, its effects on the practices of certain operators, and its bearing on other instruments. The paper concludes that the system is capable of improving environmental protection and does not pose any particular difficulties on the administrative and financial side or in international relations.

Internationally, financial security instituted by a specific agreement would be equivalent in effect to standardisation of national penalties and rules of liability as they relate to the transport of hazardous waste, without in fact amending them. The difficulties inherent in any treaty in these areas are substantial, and financial security may well provide a simpler solution to questions of penalties and liability at international level.

DESCRIPTION OF THE FINANCIAL SECURITY MECHANISM

If a waste carrier were to incur a civil penalty for failure to deliver to a disposer the waste supplied by a generator, and if the penalty were greater than the transport and disposal costs, it would be in the carrier's interest to ensure that the shipment is not mislaid. His position would then be the same as when carrying conventional freight. The penalty would convert the problem of the transport of negative-value good (that negative value being essentially the disposal cost) into the ordinary problem of transporting positive-value goods (positive value being the difference between the penalty and the cost of disposal).

To perform this function the penalty must be sufficient to deter. Assuming, for instance, that control between departure and arrival is 33 per cent effective (i.e. it detects one in three losses of hazardous waste), an effective penalty would have to be at least three times the cost of disposal, say $300 per tonne. If this were the case the penalty would very rarely be applied, because the incentive to mislay waste would have been removed.

In the event of waste being mislaid the carrier (1) would be bound to pay a penalty over to the competent authority (the Treasury, or a special environmental protection fund), under an agreement that the authorities can enforce immediately when waste is lost or not delivered to the eliminator (2). The penalty thus takes the form of payment of some form of financial security put up by the carrier to guarantee that the shipment is delivered. Where loss of the shipment was due to unforeseeable and unavoidable circumstances, payment would not be required. It would not be required either if the authorities considered that every reasonable step had been taken to avoid loss. In addition, there could be provision for the carrier to reclaim the security after payment if, for instance, he can demonstrate that he is not at fault (see section 3 below).

The security would be payable immediately without delay or legal process, simply when required by the authorities and with no possibility of any stay of execution (cf. tax law, seizure and distraint, etc.). There are various ways of achieving this end ranging from a binding undertaking to pay on the authority's demand to the provision of a deposit or surety prior to a movement. Although the effect of the security is to prevent the loss of waste and, in some degree, to compensate victims, the system is designed not to provide redress or a compensation fund but simply to prevent environmental damage.

186

Financial security has already been employed to ensure that waste is not abandoned in the environment. In Norway and Sweden car buyers lodge a deposit that is repaid with interest when the vehicle is eventually delivered to an approved breaker. In Sweden and several U.S. States a deposit on drink containers was introduced largely to prevent their careless disposal. In Switzerland a substantial charge would be levied when waste in transit does not in fact leave the country (3). Deposits are common customs requirements in many countries in relation to international road transport. Broadly speaking, financial security has long been viewed as a multipurpose economic instrument capable of strengthening statutory requirements and overcoming the weaknesses inherent in regulatory systems.

It is therefore appropriate to consider the possible role of this economic instrument in relation to transfrontier movements of hazardous waste in order to put an end to the activities of unethical operators attracted by the substantial profits to be made at the expense of the community and the environment.

FINANCIAL SECURITY IN THE INTERNATIONAL CONTEXT

Financial security could be introduced by international agreement to supplement a simple system of prior notification of transfrontier movements and consignment notes. The generator normally has to notify the disposer of his intention to send waste and has to provide a description of the waste shipment for the carrier and the disposer. The authorities further have to be notified of the movement (in line with the OECD Decision). Last, the disposer has to notify the generator that the waste has, or has not, been delivered.

In the case of a transfrontier movement of hazardous waste, financial security could be provided by the carrier (4). The security would relate to waste in the custody of the carrier in each of the countries concerned by the movement. If approved carrier status is required by these countries, the terms on which approval is given may be such that no particular difficulty arises in the event that the security has to be realised (5).

Accordingly, for every transfrontier movement the carrier would need a copy of the financial security commitment and a consignment note describing the waste. The shipment may be checked for compliance with the description at any point, and more especially at frontier crossings. There should further be a special inspection on arrival to certify that the carrier has delivered the waste to the disposer.

If the shipment were entirely mislaid before reaching its destination, the generator would not receive the waste delivery certificate (and the authorities, in addition, would not receive notification of the arrival of the waste, though they would have been notified of its departure). The generator would then be bound to notify the authorities of the countries concerned, otherwise he would become an accomplice in the misdirection of the waste (in such an event the generator and the carrier would be jointly liable, in particular for payment of the security). The carrier would be liable to prosecution in the countries concerned for failing to deliver the waste to the disposer or to report its disappearance. He would further be bound to release the security to the country designated under the agreement (for instance, the exporting country which checks outgoing waste or the country of disposal which checks whether reported shipments have in fact been delivered).

Should the shipment delivered fail to match the description in the consignment note, the security would be realised by the authority of the country where non-compliance was discovered. Should another authority prove that non-compliance was due to illegal tipping in its country (for instance, when the waste is found there), it may apply for the security to be remitted to it.

Moreover, should waste be shipped outside the area covered by the agreement, the security would be payable to the authorities of the exporting country when the authorities of the country of destination do not certify that all the waste has been received or do not permit it to be imported. In thus case the security would be a disincentive both to mislay the shipment between the countries covered by the agreement and the country of destination, and to ship hazardous waste without the express consent of the country of destination.

Last, should waste be imported from countries not covered by the agreement, the security requirement would take effect when the shipment entered the agreement area.

When abandoned waste comes from a foreign generator but the carrier is unknown, it may seem fair that the generator should pay over the appropriate security to the authorities of the country where the waste was found, as if the movement had been undertaken and notified by the generator himself. Security might even be higher (6) in this case, or coupled with a heavy fine.

The amount of security should be fixed by agreement and be freely transferable among the signatory countries. It should be easy to adjust to allow for inflation, the effectiveness of waste shipment controls, the mode of transport and disposal costs. In some cases the security could be lower, or higher. At all events the security should be determined by reference to a straightforward scale not entailing any complex analysis or calculation, which would give rise to disagreement and delay.

PAYMENT AND RETURN OF SECURITY

The amount of the security payable would be proportionate to the quantity of waste missing and would have to be paid unless the authorities considered that the loss could be justified. Afterwards the carrier could sue the person responsible for the loss. In addition he could insure himself against genuine accidental loss (7). As the insurer will not cover losses from illegal acts of the insured, the matter would be one for negociation between the carrier and his insurer in doubtful cases.

Cases where loss could be justified would be analogous with the usual heads of exoneration from liability (natural disaster, insurrection, act of war, etc.). They might extend to cases where loss is due solely to the act of an identified third party, unrelated to the entities concerned, or of a government department. The most difficult cases are undoubtedly vehicle theft and jettisoning of the shipment in an unauthorised tip, and leakage from tanker lorries (is the theft or leakage "accidental" or not?).

Security which has been paid over would be returned to the carrier if the waste were subsequently located and disposed of by the entities concerned. It might also be returned in the event of a court finding that the carrier has no liability for the loss. In this case, the amount paid back would include the interests on the amount deposited.

If repayment were to be based on no-fault liability, the carrier would incur no-fault liability up to a limit equal to the security, and conventional liability for damage in excess of the limit. The system proposed would thus enable the equivalent of a uniform liability system, up to the limit of the security, to be introduced for transfrontier movements. But the conventional liability system could also be used, with the security being returned with interests if the liability of the carrier is not established (8).

ADVANTAGES OF FINANCIAL SECURITY AS A SYSTEM

The system as proposed simply involves an undertaking to pay in favour of a given authority. The security is then realisable without delay by the authority entitled to claim it. No particular expenditure is involved in the case of most transfrontier movements, nor are there any administrative complications (9). If all the waste reaches its destination, the security has no administrative or financial costs.

Provided relatively reliable inspection is carried out on departure and arrival (10), the financial security will help to avoid losses of waste. Inspection on arrival is vital since it will identify discrepancies against the described shipment. Inspection on departure (11) is designed to ensure that the shipment does not initially involve a larger quantity of waste than shown in the description. Official inspection need not be systematic. Generators and disposers (12) may be instructed to perform these inspections systematically on their own responsibility. One advantage of financial security is hence to allow less frequent official inspection since the system is a deterrent to illegal operations.

In addition, the security helps to finance emergency measures which the authorities may be compelled to take when they find improperly tipped foreign waste ; this aspect assumes particular importance inasmuch as many years may pass before waste is discovered. But the level of the security, at $300 per tonne for instance, will be too low to cover the cost of removing and disposing of waste tipped on unauthorised sites (between $500 and $2 000 per tonne (13). If there are any pollution victims the authorities may give an advance on compensation (14).

Last, an indirect effect of the proposed system would be to keep certain operators out of the market, which would be confined to reliable transport concerns employing trained staff and sound vehicles. In that case there would be less need to introduce severe criteria for the approval of carriers and other operators between generator and disposer. In addition, generators will take care in selecting reliable carriers since they may in some cases be jointly liable with the carrier for payment of the security.

BEARING ON OTHER LEGAL INSTRUMENTS

The financial security acts to some extent as a potential heavy fine, predetermined in amount, payable on clear conditions, and the same in all countries parties to the agreement (15). It would be paid in the event of loss and returned if the waste apparently lost were subsequently located and disposed of by the enterprises concerned (16).

When waste is found, sometimes many years after it was tipped, and damage or disposal costs are incurred, the financial security will amount to a special system of liability, limited to the value of the security.

If the rule applied to repayment of security is the same as under the existing liability rules (17), the effect of the financial security system will be to transfer funds to the authorities for the time it takes to establish whether the carrier is liable or not. But if the rule applied is equivalent to no-fault liability, the financial security would, as regards compensation, apply a kind of no-fault liability limited to the amount of the security, but without prejudice to the rules of liability otherwise in force (the effectiveness of which diminishes as the parties concerned disappear, and as rights of action are extinguished by prescription).

Rather than introducing a financial security system for transfrontier movements, it would be possible to strengthen penalties and amend the rules of liability in the signatory countries by international agreement. The great advantage of the financial security system at international level is that it achieves the same outcome without having to tackle differences in national legal systems. The purpose of the security is not to replace inspection, but simply to discourage losses of waste by providing a swift and effective financial deterrent which bolsters the inspection system. It is accordingly appropriate that the security should be paid out without delay even where it is not proven that the carrier has committed an offence or is at fault. This strict approach would be counterbalanced by simple repayment rules. Carriers would no longer benefit from difficulties in obtaining proof, lengthy proceedings and low penalties, and would have to pay out the security in the event of loss before the source of the funds can disappear (e.g. short-lived firms which reappear under different names).

DISCRIMINATION BETWEEN DOMESTIC AND TRANSFRONTIER MOVEMENTS

To require security only when a shipment crosses a frontier may be seen as introducing unwarranted discrimination, within the Common Market for instance. The discrimination in question is justified by the fact that the national community alone bears the consequences of illegal tipping of hazardous waste. If this appears unwarranted to certain groups of countries,

190

there would be no obstacle to extending the financial security system to all movements of hazardous waste within those countries. Extending the scheme in this way would have the advantage of preventing losses of national waste and losses of foreign waste with the same degree of effectiveness.

CONCLUSIONS

The introduction of a system of financial security for transfrontier movements of hazardous wastes should bring about a considerable reduction in environmental damage from foreign waste and allay public misgivings. Instituted under international agreement, such straightforward deterrent would undoubtedly be more effective than a necessarily costly proliferation of regulations and controls to protect the national environment. It could prove easier to implement than internatinal reform of the liability system in respect of hazardous waste of harmonisation of penalties, but it would achieve the same results.

In addition, the financial security system would be in line with the OECD Ministerial Decision on environmental policies, under which Member Governments "...will employ economic and fiscal instruments, in combination as need be with regulatory instruments, to induce...enterprises... to anticipate the environmental consequences of their actions and take them into account in their decisions".

It accordingly seems appropriate to consider this method (the main provisions of the proposed system are summarised in the Annex) of controlling transfrontier movements of hazardous waste more effectively within the OECD area without the need for international treaties of the kind applying to nuclear waste and oil spills.

Although the system proposed is a tough and unusual one, it merits consideration in view of the major significance of hazardous waste for the economy, environmental protection, and public opinion.

NOTES AND REFERENCES

1. The term carrier covers all persons involved in shipment between the waste generator and the agreed disposer . Where more than one party is concerned (e.g. waste collector, broker, rail carrier, road carrier, ship carrier, operator of temporary storage facilities) it needs to be decided whether the security should be provided singly or jointly. Where there are reasons for thinking that the carrier may be unable to pay the security, it might have to be provided jointly by the carrier and the generator or by some approved body (for instance, a bank or a trade or industry association). Another course would be for the generator to guarantee payment of the security by the carrier, in the event of insolvency of the latter. Any of these arrangements would discourage firms from using carriers which are unreliable or have inadequate financial resources.

2. To avoid delay in payment of the security and drawn-out legal proceedings to enforce payment, it could be provided that the security would be paid automatically by some approved body if the waste delivery certificate were not received within a specified period. For example, a bank could be required without possibility of stay to transfer the security to the authorities of the disposing country one month after the departure of the waste, if the waste delivery certificate had not by then been received. This would be less expensive than a system based on the deposit of security prior to shipment.

3. In Switzerland, when goods in transit are carried in sealed trucks, a security of 50 SF/kg is usually provided. A proposal to use the same system for hazardous waste (10 SF/kg) is being studied.

4. Alternatively, security could be given by the generator since he selects the carrier, usually has greater financial resources, and is aware that the waste is hazardous. Disposers might also be involved, since shipment takes place under a contract between generator and disposer. In view of the range of legal remedies available and the variety of possible contractual provisions, it does not seem essential that the security be provided by the carrier. In the transport of nuclear waste the generator alone (without the carrier) is liable for damage, whereas in maritime transport (oil spills, for instance), the shipper is liable for part of the compensation. If a generator is financially liable for accidental discharge or tipping of hazardous waste he is likely to assist the authorities more diligently in an emergency. Co-operation of this kind, moreover, would be in line with the OECD Decision, under which the generator should "reassume responsibility for the proper management of its waste ... if arrangements for safe disposal cannot be completed".

5. One condition for approval may be a minimum share capital and the obligation for the transport firm to take out adequate insurance for the carriage of hazardous waste.

6. If security were set at $300 per tonne, this would be close to the current level for shipowners' liability for oil spill damage (133 SDRs per registered ton or about $293 per metric tonne under the 1969 Brussels Convention and 420 SDR per registered ton or $ 924 per metric tonne under the 1984 Protocol) or the level of carrier's liability for loss resulting from loss of or damage to goods (1.5 SDR per kilogramme) under the United Nations Convention on the carriage of goods by sea (Hamburg, 1978). It would seem difficult to increase the security appreciably above $300 per tonne. On the other hand, the security must not be less than disposal and transport costs.

7. It would be of value for the carrier's insurance substantially to exclude the consequences of wrongful acts by the generator or employees of the carrier so as to ensure that the financial security system is an effective deterrent.

8. It could also be agreed that emergency measures which governments take against improper discharge should in any case be financed from the security (liability for expenses of public bodies). In other words, in line with the polluter-pays principle, these measures would not be charged to the taxpayer and the carrier would have to recover the costs of the emergency measures from the person responsible for the improper discharge.

192

9. The system as proposed can operate only with proper checks that shipments have been delivered. It has negative financial consequences for the carrier since he may in some cases have to pay out the security prior to any decision by a criminal, administrative or civil court. However, interest on the security will be credited to him when the security is paid back. There may also be cases where a carrier will pay out the security without being solely responsible for the loss (for instance, losses caused in part by insolvent or unidentified persons).

10. Where the system provides for automatic payment of security unless a delivery certificate is received (see footnote 2 above), absence of the certificate will trigger off payment of the security and intervention by the authorities. Verification of consignment documents, normally carried out by the authorities, would in this case be undertaken by the body authorized to provide the security. This would be more expensive since it would entail additional administrative work by the body concerned.

11. The system of financial security would not preclude consignments being sent under a false description by the generator, in particular those sent abroad or onto the high seas. If the proposed system proves effective, one indirect outcome could be to boost the fraudulent practice whereby certain materials are not regarded as hazardous waste. To discourage this practice it seems appropriate that the carrier of a misdescribed consignment should be obliged to pay out the security immediately before continuing the journey, and should recover the security only when the waste has been legally disposed of. Return of the security could also be deferred until the carrier had paid fines imposed for concealing the true nature of the waste and participating in an illegal transfrontier shipment. In such circumstances the generator will no doubt be jointly responsible with the carrier for payment of the security and fines.

12. A disposer may be unwilling to check shipments received or report discrepancies for the sake of good business relations, and it may be necessary for inspection to be carried out by an independent body. Thorough inspection might be encouraged by means of a bonus for identifying shipments not in accordance with the consignment note. Bonuses of this kind are payable to tax, customs and police bodies in France. Prison accounts officers, for instance, are entitled to a bonus of 2.5 per cent of payments made on behalf of prisoners serving sentences in respect of sums due to central or local government authorities (Section D 322 of the Code of Penal Procedure, Act of 7th March 1975). The practice of bonuses is not considered acceptable in some countries, which prefer to apply severe penalties. As a general rule, all parties involved should be required to give undertakings concerning the proper and trouble free conduct of operations so as to discourage unlawful behaviour. (In this connection, see the declarations required in France under the Order of 5th July 1983 concerning the importation of hazardous and toxic waste).

13. The cost of cleaning up a site where hazardous waste has been tipped may be up to $2,000 per tonne of hazardous waste (figure supplied by the US Environmental Protection Agency, quoted by Congressman J. Florio, "Congressional Involvement in Hazardous Waste", Consulting Engineer,

p. 44, March 1984). With cleaning-up costs estimated to exceed $500 per tonne, security of $300 per tonne would be insufficient to cover liability but might be enough to finance emergency measures.

14. The problems raised by transfrontier movements of hazardous waste are different from those involved in oil shipment by sea since the chief aim in the first case is to prevent loss of the waste and the ensuing cost of restoring the environment, whereas in the case of oil the chief purpose is to compensate victims after an accident . Insurance and a compensation fund will ensure that victims are indemnified in the event of damage, but are inoperative until the damage actually occurs. Mechanisms along these lines to pay for the removal of waste from illicit tipping (the Superfund in the United States, for instance) help to spread the financial burden on waste generators, but would hardly be a deterrent to unscrupulous practices. Since the financial security is likely to be less than the cost of removing and eliminating waste after illicit tipping, a system of this kind would cut down the frequency of such tipping but would leave untouched the problems associated with cleaning them up.

15. In the case of customs fraud, penalties include fines and confiscation of the goods. Waste being of little or no value, confiscation would have no effect. A fictitious value liable to confiscation, in the form of the financial security would be useful. This clearly demonstrates that the security is not exactly comparable to a fine, even though its function is similar. When used in conjunction with recyclable materials which can also be considered as hazardous waste, the financial security would provide a very positive value to materials whose value can be positive or negative according to the conditions of the market. If carriage of this materials would be subject to less strict control measures, financial security would reinforce the weaker control system.

16. Under the French Act of 19th July 1976 (Section 23), the operator of a classified installation may be required to "lodge with a public accounting officer a sum to cover the cost of the work to be carried out, to be returned to the operator as the work proceeds". The financial security performs the same function, but is paid over by carriers whose waste disappeared outside an approved disposal site.

17. Where the security is paid by an approved body (see footnotes 2 and 9 above) the agreement might usefully specify the terms on which the authorities will return the security. To ensure international uniformity, identical repayment conditions could be adopted by all countries (for example, repayment when the carrier establishes that he is not at fault, or in cases of exoneration from no-fault liability only). Another course would be for the security to be repaid in cases where the carrier is not liable for the loss under the law applicable to domestic movements of hazardous waste within the country where the security was paid.

Annex

SUMMARY OF THE MAIN PROVISIONS OF AN INTERNATIONAL SYSTEM
OF FINANCIAL SECURITY FOR TRANSFRONTIER MOVEMENTS
OF HAZARDOUS WASTE

(a) For any transfrontier movement of hazardous waste the approved carrier undertake to pay out security if he does not deliver the waste which he has received from a generator to the disposer, and to report forthwith any loss of waste en route.

(b) The authorities in the country of disposal (or their agents) check that the waste received corresponds to the waste described in the consignment note.

(c) In the event of loss of hazardous waste en route, the authorities in the country where the loss is detected must notify the authorities in the other countries concerned and call in the security immediately.

(d) Payment of security does not release the generator or the carrier from any obligations they may have in respect of the hazardous waste.

(e) If one of the countries concerned can subsequently establish that the hazardous waste was lost while crossing its territory, it will receive the security from the country to which it was paid, or will if necessary itself call in the security.

(f) Where hazardous waste leaves the area covered by the agreement for disposal in a third country, the security is paid to the generator's country if the latter does not receive the necessary certificate from the importing country.

(g) Where hazardous waste enters the area covered by the agreement, security must be provided by the carrier.

(h) The generator of hazardous waste which is found abandoned in another country must pay that country the appropriate security, unless he can prove that it has already been paid. He will also have to pay the security if undeclared hazardous waste generated by him is found in another country.

(i) The eliminator should pay the security if he falsely declares that he has received hazardous waste or fails to report that hazardous wastes was not delivered to him. .

(j) The security will be repaid when lost waste is ultimately disposed of, subject to any costs incurred by the authorities to ensure that the waste does not constitute a threat to the environment.

(k) The security will also be repaid in other circumstances, to be specified. The carrier may take proceedings for return of his security in the courts of the country to which the security was paid.

(l) The carrier, the generator and the eliminator may take proceedings against any person whose acts or omissions have resulted in payment of security.

(m) The amount of the security will be fixed by agreement among the parties, who will also provide for a simple procedure for reviewing such amount. The security is to be freely transferable from one country to another.

(n) The parties may by agreement exclude certain hazardous waste from the financial security system or adjust the obligation according to the type of waste, the mode of transport, etc.

(o) Disputes among parties concerning the application of the agreement are to be settled by negotiation or arbitration.

Note: The parties will have to decide the extent to which generators and disposers may be covered by the financial security system, and the circumstances in which the security is to be returned. They might also adopt some simple and automatic way of making the security available in all cases (bonds, etc.).

TRANSPORT OF HAZARDOUS WASTE

Alan I. Roberts*
Department of Transport, Washington, United States

SUMMARY

International controls (laws, regulations, conventions, agreements, etc.) for transport of hazardous wastes should be based upon existing controls for transport of dangerous goods. The Recommendations of the United Nations Committee of Experts on the Transport of Dangerous Goods are being implemented on a worldwide basis. The most dangerous of hazardous wastes are already subject to existing transport regulatory controls. Development of new and different transport controls for hazardous wastes, except for special documentation, may be counterproductive to effective control of hazardous waste transport.

INTRODUCTION

Appropriate controls for the transport of hazardous wastes are essential to the implementation of an international hazardous waste disposal programme. While there is agreement that controls are necessary, the nature of those controls and the methods for their implementation are not as yet decided on a worldwide basis. This paper addresses transport measures. It does not address in any manner the methods that should be used for storage, treatment, or disposal of hazardous wastes or whether a waste should be permitted to be transported into a State for such a purpose. Once those decisions are made and transport is to occur, problems concerning transport may arise. We must be certain those problems are minimized to assure public safety and proper delivery to authorized facilities.

---

* Associate Director for Hazardous Materials Regulations of the Materials Transportation Bureau (MTB), Research and Special Programs Administration.

In the following, I present five conclusions and discussions thereof concerning transport of hazardous wastes. They are based upon my knowledge and experience and are not intended to represent a position of the United States Government or the United Nations Committee of Experts on the Transport of Dangerous Goods of which I am current Chairman.

1. The United Nations Recommendations on the Transport of Dangerous Goods are being implemented on a worldwide basis by international transportation organizations and by a number of UN member States

The United Nations Committee of Experts on the Transport of Dangerous Goods (hereafter referred to as the "UN Committee") was established by a resolution of the United Nations Economic and Social Council on April 15, 1953, [ECOSOC Resolution 468 G (XV)] and first met in 1954. The terms of reference for the work of the UN Committee, which have remained fundamentally unchanged over the years, are:

(i) Recommending and defining groupings or classification of dangerous goods on the basis of the character of risk involved;

(ii) Listing the principal dangerous goods moving in commerce and assigning each to its proper grouping or classification;

(iii) Recommending marks or labels for each grouping or classification which shall identify the risk graphically and without regard to printed text; and

(iv) Recommending the simplest possible requirements for shipping papers covering dangerous goods.

At the time of the establishment of the UN Committee there were three basic codes governing the carriage of dangerous goods in international commerce. The first was the International Regulations Concerning the Carriage of Dangerous Goods by Rail (RID) which are included in Annex 1 of the International Convention Concerning the Carriage of Goods by Rail (CIM), established in 1890. The second was the regulations of the Interstate Commerce Commission (ICC) in the United States which were first issued in 1908 and were used by countries in North America and largely reproduced by the International Air Transport Association (IATA) in its restricted articles regulations. The third was the British "Blue Book" which was broadly recognized as the basic standard for the international transport of dangerous goods by sea. Each of these codes was based on a different approach relative to classification and labeling of dangerous goods.

On 30 September 1957, before recognition of the UN Recommendations by any official body, the European Agreement Concerning the International Carriage of Dangerous Goods by Road (ADR) was concluded under the auspices of the Economic Commission for Europe (ECE). These regulations largely reproduced the provisions of the RID regulations in order to facilitate intermodal transport within Europe. This action is inderstandable in light of the strong desire for harmony between the rail and highway modes within Europe and because the UN Recommendations were not yet fully developed.

In 1965, a significant action was taken by the International Maritime Consultative Assembly in Paris when it adopted the International Maritime Dangerous Goods Code. The Code was based primarily upon recommendations of the UN Committee. Since adherence to international maritime rules is necessary for any nation involved in maritime trade, the IMCO action stimulated increased interest in the work of the UN Committee and its efforts to create harmonization of international standards for the transport of dangerous goods.

More recently, the International Civil Aviation Organizations's (ICAO) Dangerous Goods Panel has developed a complete set of regulatory standards for air transport of dangerous goods. These standards are based almost entirely upon recommendations of the UN Committee and became effective on 1 January 1984.

Clearly then, there are a number of international organizations engaged in the development and promulgation of standards governing the international transport of dangerous goods. These standards may have worldwide application to only one mode of transport, or even to a single mode of transport within a particular region of the world. However, they are increasingly being revised to reflect the recommendations of the UN Committee in order to provide the intermodal/interregional harmonization essential to the smooth and efficient transport of goods under the modern transport system and, consequently, during the 30 years following the establishment of the UN Committee, a considerable degree of harmonization of international standards has evolved.

A brief description of the various international organizations involved with the development of dangerous goods transport standards, as well as an explanation of the relationships between these organizations, follows:

UN Committee of Experts on the Transport of Dangerous Goods

The United Nations Committee of Experts on the Transport of Dangerous Goods is responsible for the development of recommendations dealing with the multimodal transport of dangerous goods, including provisions for the classification, labeling, and packaging of these materials. The Committee, which reports directly to the UN Economic and Social Council (ECOSOC), is currently composed of members from Canada, France, Federal Republic of Germany, Italy, Japan, Norway, Poland, the United Kingdom, the United States of America, and the Soviet Union. The Recommendations of the Committee are published by the UN in a volume entitled Transport of Dangerous Goods. ECOSOC, by resolution, has urged all member States, regional economic commissions and international organizations to bring their practices for the transport of dangerous goods into conformity with the UN Recommendations. As already mentioned, significant progress has been made toward this end with the decisions by the two major intergovernmental organizations responsible for development of recommendations dealing with transport of dangerous goods by water (IMO) and air (ICAO) that the UN Recommendations would serve as the framework for their recommendations. In addition, considerable progress has recently been made toward the harmonization of certain regional modal requirements with the UN Recommendations, such as those of the ADR and the RID.

The Committee of Experts is composed of two subsidiary bodies, the Group of Experts on Explosives and the Group of Rapporteurs on the Transport of Dangerous Goods. The Committee meets biennially to consider the reports of the subsidiary bodies and forwards its suggested amendments to the Recommendations to ECOSOC for final approval.

UN Group of Experts on Explosives

The Group of Experts on Explosives is one of the subsidiary bodies of the UN Committee of Experts on the Transport of Dangerous Goods. The Group is responsible for making recommendations to the Committee relative to the transport of explosives by any mode and to prepare amendments to the UN Recommendations as necessary for consideration by the Committee of Experts. The work of the Group typically involves the consideration of the suitability of new explosives articles or substances for listing in the UN Recommendations as well as the necessary packaging and labeling requirements for these materials and test methods and criteria for the classification of explosives. The work of this Group has a considerable influence not only on the transport of explosives but also on the handling of explosives. For example, several military organizations have adopted the UN system as the basis for requirements for the storage and handling of explosives. The Group of Experts on Explosives formally meets, in accordance with the directives of the Committee, once a year and informal special meetings are held when special problems need to be addressed.

UN Group of Rapporteurs on the Transport of Dangerous Goods

The Group of Rapporteurs on the Transport of Dangerous Goods is the second of the subsidiary bodies of the United Nations Committee of Experts on the Transport of Dangerous Goods. The Group is responsible for making recommendations to the Committee relative to the multimodal transport of all dangerous goods, with the exception of explosives, and to prepare any necessary additions or amendments to the UN Recommendations concerning these materials for consideration by the Committee. The scope of work of this group is broad, including the development of recommendations relative to the listing, classification, labeling, placarding, packaging, consignment procedures (including transport documentation), and design and construction standards for tank-containers.

The Group of Rapporteurs meets in accordance with the directives of the Committee of Experts, normally three times during a two-year period for a duration of one or two weeks at each meeting. In addition, due to the nature and complexity of the work, and because of the vast quantity of work remaining to be accomplished, intersessional meetings are sometimes found necessary.

IMO Subcommittee on the Carriage of Dangerous Goods (CDG)

The CDG Subcommittee is one of a number of subcommittees of the Intergovernmental Maritime Consultative Organization's Maritime Safety Committee. The CDG Subcommittee is responsible for the development of recommendations dealing with the transport of packaged and bulk solid dangerous goods by sea. The Subcommittee has published a comprehensive set of recommendations known as the International Maritime Dangerous Goods (IMDG) Code which deals with all aspects of the transport of packaged dangerous goods by sea (i.e., packaging, marking, labeling, classification, documentation, placarding, stowage, segregation and handling of dangerous goods). The basic systems of listing, classification, labeling and packaging appearing in the IMDG Code are derived from the recommendations of the UN Committee of Experts on the Transport of Dangerous Goods.

The IMDG Code is widely recognized throughout the world as the minimum standard for transport of dangerous goods by sea. It has been fully adopted into the regulations of at least 34 Nations and is in various stages of adoption in many more.

ICAO Dangerous Goods Panel (DGP)

The Dangerous Goods Panel (DGP) is one of the subsidiary bodies of the International Civil Aviation Organization's (ICAO) Air Navigation Commission. The DGP, which first met in January 1977, prepared an Annex to the Convention on International Civil Aviation (the Chicago Convention) which, for the first time, established internationally agreed upon intergovernmental standards for the safe transport of dangerous goods by air. This Annex is supported by a detailed body of Technical Instructions, which are incorporated by reference into the Annex and provide the necessary details concerning classification, packaging, marking, labeling and handling aboard aircraft of dangerous goods shipments. Both the Annex and Technical Instructions are based upon the recommendations of the UN Committee of Experts and the International Atomic Energy Agency (IAEA), thereby providing much needed harmonization of the international transport requirements for dangerous goods by air with those of other modes.

The ICAO Council adopted the Annex and Technical Instructions in June 1981, and set January 1984 as the mandatory compliance date for implementation of these standards.

European Agreement Concerning the International Carriage of Dangerous Goods by Road (ADR)

The European Agreement Concerning the International Carriage of Dangerous Goods by Road (ADR) was first published in 1959. It consists basically of three sections. The first section contains the actual Agreement and protocol of signature. The second section, Annex A to the Agreement, contains the provisions concerning dangerous substances and articles; that is, the list of substances allowed to be carried as well as the requirements for labeling and packaging of the materials. The third section, Annex B to the Agreement, sets forth the operational requirements for carriage by road as well as the requirements for transport vehicles, cargo tanks and portable tanks transporting dangerous goods.

The ADR is administered through the Inland Transport Committee of the United Nations Economic Commission for Europe. Over the past several years, the primary thrust of amendments to the ADR has been to incorporate the Recommendations of the UN Committee, and amendments incorporating substantial portions of the UN Recommendations will become effective on 1 January 1985.

International Regulations Concerning the Carriage of Dangerous Goods by Rail (RID)

The International Regulations Concerning the Carriage of Dangerous Goods by Rail (RID) are found at Annex 1 to the International Convention Concerning the Carriage of Goods by Rail (CIM). CIM is an international convention; however, it is not of worldwide applicability since its provisions apply to only the European Nations which are signatory. The convention is administered by the Central Office of International Rail Transport (OCTI) in Berne, Switzerland. OCTI is not associated with the United Nations. However, RID and ADR conduct joint meetings frequently in an ongoing attempt to maintain harmony between the conventions governing road and rail transport in Europe. The majority of the work of the Joint meetings in recent years has been directed toward the introduction of the UN Recommendations into both the RID and ADR. Consequently, as is the case with the ADR, amendments to the RID incorporating substantial portions of the UN Recommendations will become effective on January 1, 1985.

International Atomic Energy Agency (IAEA)

The International Atomic Energy Agency (IAEA) came into existence on 29 July 1957, as a result of an international conference convened at the United Nations Headquarters. The IAEA has published regulations for the safe transport of radioactive materials since shortly after its establishment. However, in 1977 the Secretary General of IAEA, in recognition of the acute public awareness of the transport of radioactive materials, established the Standing Advisory Group on the Safe Transport of Radioactive Materials (SAGSTRAM), in order to maintain these rules current in light of technological advancement.

The IAEA rules, known more commonly as Safety Series 6, are used extensively throughout the world as the basis for regulations governing the safe transport of radioactive materials. They have been incorporated into the U.S. DOT's Hazardous Materials Regulations as well as the regulations of many other Nations. In addition, they have been adopted into various international standards for the transport of dangerous goods including the IMO IMDG Code, ICAO Technical Instructions and the RID and ADR Agreements.

In summary, during the years following the initial IMO recognition of the UN Recommendations, a number of actions have been taken by member States and international bodies relative to harmonization under the UN system. Some of the more significant of these include: (1) The UN listing of dangerous goods with associated UN identification numbers, (2) Standardized labels for identification of dangerous goods, (3) Defining criteria for the classification of dangerous goods, (4) Standardized documentation in terms of the method of display of dangerous goods information, (5) Packaging performance standards, (6) Standard design requirements for intermodal tank-containers, and (7) Specialized technical requirements pertaining to materials such as explosives and organic peroxides.

2. Existing transport regulatory systems should be used for implementation of hazardous waste control programmes

The principal source of transport services for movement of hazardous wastes will be companies already experienced in the transport of dangerous goods. In my more than 25 years experience in the transportation field, one basic theme has prevailed relative to regulatory requirements affecting transport companies and their employees. The regulatory system must be relatively simple, otherwise our expectations for compliance by transport workers should not be very high. Most transport workers do not have a university education. Our instructions to them must be dictated by their ability to comprehend if we expect our instructions (regulations) to be followed. To establish a distinct and separate regulatory system for control of the transfrontier movement of wastes would, in all likelihood, be ineffective in accomplishment of our goal which is safe and responsible transport of these wastes.

For two years, I served as Vice President of a trucking company in the United States. During that experience, one of my responsibilities was training the drivers employed by the company. I would estimate that more than 75 percent of the training given our drivers addressed safety in the operation of our vehicles, including compliance with Federal safety requirements. The United States Government requires drivers used in the transportation of dangerous goods to be trained by the companies employing them concerning the Federal requirements for transport of dangerous goods. Penalty actions have been taken against companies who have failed to train their drivers. The greatest complaint raised by companies subject to the requirements relates to the complexity of our regulations. Since transport personnel are already required to be trained and are becoming knowledgeable concerning international dangerous goods transport requirements, it would be appropriate to base hazardous waste transport requirements upon those requirements in order to avoid greater complexity. In the United States this has been accomplished by a close working relationship between the U.S. Department of Transportation (DOT) and the U.S. Environmental Protection Agency (EPA). In fact, there was an agreement that transport rules of DOT would be incorporated by EPA without duplication in EPA rules and vice versa. In order to illustrate this relationship, the following is quoted from a note introducing 40 Code of Federal Regulations, Part 263 which contains EPA's standards applicable to transporters of hazardous wastes:

Note: The regulations set forth in Parts 262 and 263 establish the responsibilities of generators and transporters of hazardous waste in the handling, transportation, and management of that waste. In these regulations, EPA has expressly adopted certain regulations of the Department of Transportation (DOT) governing the transportation of hazardous materials. These regulations concern, among other things, labeling, marking, placarding, using proper containers, and reporting discharges. EPA has expressly adopted these regulations in order to satisfy its statutory obligation to promulgate regulations which are necessary to protect human health and the environment in the transportation of hazardous waste. EPA's adoption of these DOT regulations ensures consistency with the requirements of DOT and thus avoids the establishment of duplicative or conflicting requirements with respect to these matters. These EPA regulations which apply to both interstate and intrastate transportation of hazardous waste are enforceable by EPA.

DOT has revised its hazardous materials transportation regulations in order to encompass the transportation of hazardous waste and to regulate intrastate, as well as interstate, transportation of hazardous waste. Transporters of hazardous waste are cautioned that DOT's regulations are fully applicable to their activities and enforceable by DOT. These DOT regulations are codified in Title 49, Code of Federal Regulations, Subchapter C.

EPA and DOT worked together to develop standards for transporters of hazardous waste in order to avoid conflicting requirements. Except for transporters of bulk shipments of hazardous waste by water, a transporter who meets all applicable requirements of 49 CFR Parts 171 through 179 and the requirements of 40 CFR 263.11 and 263.31 will be deemed in compliance with this part. Regardless of DOT's action, EPA retains the authority to enforce these regulations.

Since the EPA established the criteria for defining hazardous wastes and specified information requirements for hazardous waste manifests, in addition to those normally specified for dangerous goods documentation, DOT's requirements implementing hazardous waste transport requirements were rather brief and easy to comprehend. They are found in DOT's regulations in 49 Code of Federal Regulations, Section 171.3, as follows:

§ 171.3 Hazardous waste.

a) No person may offer for transportation or transport a hazardous waste (as defined in § 171.8 of this subchapter) in interstate or intrastate commerce except in accordance with the requirements of this subchapter.

b) No person may accept for transportation, transport, or deliver a hazardous waste for which a manifest is required unless that person:

   1. has marked each motor vehicle used to transport hazardous waste in accordance with § 397.21 or § 1058.2 of this title even though placards may not be required;

   2. complies with the requirements for manifests set forth in § 172.205 of this subchapter; and

   3. delivers, as designated on the manifest by the generator, the entire quantity of the waste received from the generator or a transporter to:

      i) The designated facility or, if not possible, to the designated alternate facility;

      ii) The designated subsequent carrier; or

      iii) A designated place outside the United States.

      Note: Federal law specifies penalties up to $25,000 fine and 5 years imprisonment for the willful discharge of hazardous waste at other than designated facilities. 49 U.S.C.1809.

The United States began implementing UN standards in 1970 and now recognizes the UN labeling system, shipping descriptions, identification numbers, and tank-container design and construction standards and has in progress regulatory actions pertaining to UN packaging performance standards. Similar actions are presently underway in Canada. The UN standards adopted at the present time in the United States are an integral part of a hazardous waste control system. I recommend they be considered on a similar basis for implementation of regional and worldwide control systems for transport of hazardous waste rather than creation of an entirely new system.

3. A majority of hazardous wastes presenting the greatest danger are also dangerous goods already subject to transport regulation and those wastes that are not presently regulated could be included in the dangerous goods transport control system

During the past 10 years there has been a large amount of rhetoric on issues concerning hazardous wastes, including their transportation. Citizens become alarmed when they become aware that hazardous wastes are being transported in proximity to where they live and work. Setting the emotions generated by the subject aside, I believe it to be important that we recognize that, aside from the lack of incentive to deliver goods that have little or no economic value to proper and lawful destinations, it should be recognized that most hazardous wastes pose less danger during transport than the virgin dangerous goods of their origin. For example, I believe most would agree that spent sulfuric acid would pose less danger during transport than the high-strength sulfuric acid used in the manufacturing process that resulted in its being spent. Therefore, I believe a majority of the hazardous wastes presenting the greatest danger are already subject to transport regulations, including those issued in accordance with the United Nations Recommendations.

An examination of Chapter 2 of the United Nations Recommendations will disclose that it is a comprehensive listing of dangerous goods. True, it does not list all dangerous goods by technical name. However, all materials meeting the defining criteria of the UN fall within a description in the chapter. Many entries are followed by the letters n.o.s. (meaning "not otherwise specified"). For example, if a solid material meets the definition of a corrosive substance according to Chapter 8 of the UN Recommendations, and it is not listed by technical name in the list in Chapter 2, it will be described and transported in accordance with the provisions for the entry, "corrosive solid n.o.s. (UN No.1759)".

Under the UN system, all dangerous goods are classified into one of nine hazard classes. These classes are:

Class 1 - Explosives

Class 2 - Gases: compressed, liquefied or dissolved under pressure

Class 3 - Inflammable liquids

Class 4 - Inflammable solids; spontaneously combustible substances; and substances which emit flammable gases on contact with water

Class 5 - Oxidizing substances

Class 6 - Poisonous (toxic) and infectious substances

Class 7 - Radioactive substances

Class 8 - Corrosives

Class 9 - Miscellaneous dangerous substances

Some of these classes (e.g., Class 4) are further subdivided into divisions (i.e., Divisions 4.1, 4.2, and 4.3) for the purposes of a more succinct communication of the risks of a substance through use of distinctive labels for the individual divisions, and through use of the division numbers themselves in transport documentation.

The hazards of materials included within these classes are clear from the name of the class itself, with the possible exception of Class 9. Class 9 contains substances which during transport present a danger not covered by the other eight classes. For example, polychlorinated biphenyls are assigned to Class 9 (UN No.2315). In the United States, we have listed in our regulations, in a class corresponding to UN Class 9, an entry "Hazardous Waste n.o.s.". In this connection, if the UN Committee were persuaded to provide a similar entry in its Class 9, no other action would have to be taken to implement waste requirements except when it is determined that a particular substance, such as polychlorinated biphenyl, should be considered by the UN Committee for inclusion in the list. If use of a generic entry alone is not satisfactory, a provision could be added to the transport documentation requirements specifying that the technical name of the substance, or technical names of constituents in the substance, described by the generic entry must be added as part of the shipping description for the waste. This practice would be similar to the requirement currently applied to all dangerous goods being transported under an "n.o.s." entry in accordance with Regulation 4 of Chapter 7 of the 1974 Safety of Life at Sea Convention and a requirement we have in the United States for environmentally hazardous substances.

I recognize that scientific experts who address themselves to problems involving disposal of hazardous wastes and control of environmentally dangerous substances would express critical views concerning the listings of dangerous goods in the UN Recommendations. And, rightfully so in regard to the kinds of problems they must address. However, these views must be carefully weighed against the benefits to be derived in implementing a control system for transport that harmonizes with the dangerous goods transport control system. As indicated earlier, several minor adjustments could be made to the UN system that would overcome the criticisms of those experts -- in particular, in addition of a provision for additional descriptive information following "n.o.s." entries.

4. Uniform documentation for international movement of hazardous wastes, taking into account the needs of individual States, is essential to the accomplishment of an adequate hazardous waste control system

Proper documentation is fundamental to the accomplishment of any dangerous goods transport safety programme. A considerable amount of time and effort has been expended on this subject by all the international organizations involved in dangerous goods transport as well as many of the member States. In the United States, our approach to the present time has been to specify the information that must be contained on shipping papers without specifying a specific form for placement of the information. Standard to all the documentation systems for dangerous goods at the present time is display of the proper shipping name, class, and United Nations Identification Number.

At its fourth session in September 1975, the UN/ECE Working Party on Facilitation of International Trade Procedures included in its list of priorities a study of the documentary aspects of the international transport of dangerous goods. At its session in February 1978, the Working Party completed its work and submitted its recommendation No.11 (TRADE/WP.4/INF.53) to the ECE. Most of the recommendations of the Working Party were adopted by the UN Committee at its December 1978 session and are now found in Chapter 13 of the UN Recommendations.

On 18 December 1978, EPA and DOT issued their first proposals for implementation of documentation requirements addressed, in particular, to hazardous waste transport. Based upon a number of comments received from industry, EPA decided not to adopt a final regulation containing a specific manifest form. Instead, EPA and DOT adopted requirements concerning the information to be placed on hazardous wastes manifests when the final regulations were issued in early 1980. Shortly thereafter, a number of problems began to arise when individual States in the United States began to specify their own forms that were required for display of hazardous waste information. By the end of 1980, there were no less than 20 different State manifest form requirements and this began to impact upon both generators and transporters of hazardous wastes. Visualize the confusion that would prevail if each country within a region of the world were to adopt its own and distinctly different hazardous waste manifest form and require its presence whenever a waste is transported across its borders. The result would be a severe imposition on generators who would be required to go through the costly and inefficient procedure of filling out several manifest forms with duplicative information in order to ensure that their wastes reach designated disposal facilities without delay, rejection or penalty. Also, a lack of uniform information requirements could prevent generators of wastes with sites in more than one State from standardizing their manifesting procedures, a situation that would make it extremely difficult for them to implement standardized data handling programmes for coordinating their hazardous waste management programmes.

In recognition of these problems, EPA and DOT issued a new rule proposal for a uniform hazardous waste manifest and I am pleased to say that the final rule was issued on 20 March 1984. The United States' Uniform Hazardous Waste Manifest* will be mandatory nationwide on 20 September 1984. Most of the entries on the form are standardized and applied nationwide in the same fashion. However, States are permitted to a certain extent to require their own specialized information on the Uniform Manifest in the lettered blocks that are shaded. Precedence is given to the State in which the waste is to be disposed. If that State has no specialized requirements, then the State in which the waste is generated takes precedence. This is the first time that DOT has found it necessary to participate in requiring a specified form for documentation of dangerous goods. We found this action necessary to assure that wastes are not unduly delayed in transit due to paperwork requirements.

207

While the UN Committee has not expended much effort specifically addressing hazardous waste transport issues, one action of the Committee is worthy of note. At its December 1982 meeting, the Committee adopted a new provision in Chapter 13 of its Recommendations which reads as follows:

> 13.6.1.2. If waste dangerous goods (other than radioactive wastes) are being transported for disposal, or processing for disposal, the proper shipping name should be preceded by the word "Waste".

If this provision is adopted by international bodies and member States, as was recently implemented in the United States, shipping documents would be required to contain the word "Waste" as part of each dangerous goods description. This would lead to early implementation of a communication indicating the presence of wastes when documents are examined by cognizant authorities.

Since documentation would be a critical element in the implementation of a hazardous waste transport control programme, I believe the UN Committee should address the subject in greater detail than it has in the past. I am certain that other international organizations, and agencies of member States having representation on the UN Committee, could convince its members to do so.

5. A major problem to be overcome is coordination and delineation of regulatory authority between agencies within States, between agencies of States and international organizations, and between international organizations

In my view, this is the most important of my conclusions --and the most difficult for me to express in a constructive manner. Perhaps the best way to begin this discussion is to provide some background concerning our experience in the United States.

Prior to 1975, transport regulations pertaining to dangerous goods were issued separately by four different agencies, the Federal Highway Administration, the Federal Railroad Administration, the Federal Aviation Administration, and the United States Coast Guard. The role of our predecessor office was to coordinate the regulatory activities of these four agencies, in particular, those requirements pertaining to the preparation and offering of shipments for transportation. However, the predecessor office held no regulatory authority. The system did not work very well when four different agencies were required to sign (issue) each rule that had an effect on their individual modes of transportation. Recognizing that there was a need for harmonization and for one organizational entity to be accountable for all aspects of dangerous goods regulatory requirements (except for shipments in bulk by vessel), the United States Congress enacted the Hazardous Materials Transportation Act (HMTA) in 1974. The HMTA vested all regulatory authority in the Secretary of Transportation who established the Materials Transportation Bureau (MTB) shortly after the statute was enacted. By direction, The Secretary vested in MTB all regulatory authority relative to safe transportation of dangerous goods including authority to issue

---

* A copy of this Manifest is provided in the report by F. Van Veen in this book.

regulations pertaining to carrier operations. Therefore, MTB was able to address all modes of transportation when it entered into negotiations with the EPA concerning transportation of hazardous wastes.

I believe the best way for me to raise interest in my conclusion under this topic is to ask several questions for self examination within States and international organizations:

1. Does more than one agency within your government hold authority to regulate transport of dangerous goods (e.g., one for road and another for rail transport)?

2. Do different agencies in your government regulate transport of dangerous goods and hazardous wastes?

3. If your answer to 1 or 2 is yes, does one agency of your government have precedence in representing your government at international bodies that issue or plan to issue standards for transport of dangerous goods and hazardous wastes?

4. Does your government have a policy position as to whether hazardous waste transport control matters should be harmonized on the basis of regionally established standards or those established as worldwide standards?

5. Does your government have a policy position as to which international organization should establish appropriate standards for transport of hazardous wastes?

In our Emergency Response Guidebook (ERG) for Hazardous Materials (more than 1,000,000 copies have been distributed to emergency services in the United States to date), emphasis is placed on the theme "Who's in Charge". Most experts agree that establishment and acknowledgement of a command and control procedure is a major component of a successful emergency response programme. I strongly believe that a properly structured command and control programme is also necessary for a successful hazardous waste transport control programme, both nationally and internationally. Put simply --who's in charge? Industry, government, and more importantly, our citizens will benefit if we answer this question at all levels in a timely and appropriate fashion.

Author's Special Note:

In the document "Draft Guidelines for the Environmentally Sound Management of Hazardous Wastes" issued by the United Nations Environment Programme 15th December 1983 (UNEP/WG.95/4), Guideline 37 of the Draft addresses existing transport conventions, and the commentary on the Guideline states, in part, the following:

...the existing system of international carriage conventions on dangerous goods provide an extremely efficient and effective framework into which transport of dangerous goods waste should be fitted, thus enabling advantage to be taken of the existing structures governing transport of dangerous cargoes and of the specialist expertise available to the responsible agencies, such as the International Maritime Organization, overseeing these activities.

THE CARRIAGE OF HAZARDOUS WASTE AND THE LIABILITY QUESTION

Martine Rémond,
Professeur at Strasbourg University, France

SUMMARY

The purpose of this report, which analyses liability incurred in connection with the carriage of hazardous wastes under the legislation of the OECD countries, is to identify any gaps in relevant national or international law and any defects in the links between them, together with the most appropriate machinery to prevent fraud and ensure proper compensation for victims. The transport operation itself and any ancillary operations are looked at in turn from this standpoint.

INTRODUCTION

1. The specific nature of the carriage of hazardous waste

The movement of hazardous waste, whether across frontiers or not, means carriage from place to place, i.e. transport. At first sight, therefore, it would seem sufficient to refer to the tried and tested rules applying to transport in general in order to solve the questions of liability bound up with this type of transport. For transfrontier movement, the international agreements governing transport by rail, air, sea and road ought, on the face of it, to be applicable (1). But moving hazardous waste is not a "run-of-the-mill" transport operation: its safety is of more concern to others and to the environment than to the consignor and consignee. If an accident happens, this is, of course, the carrier's responsibilty but its seriousness, due to the nature of the cargo being carried, depends on the activity of the generator of the waste. In general terms, the carriage of hazardous waste cannot be qualified as an independent operation in economic terms. Closely dependent on the production of the waste and its disposal, transport here is no more than one part of the overall disposal operation from which it cannot be dissociated.

It is first necessary, therefore, to set down the specific characteristics of that operation which make it necessary for special transport rules to apply. That is the purpose of the first part of this report. Then a brief analysis is given of the mechanisms of transport law generally applicable in the OECD countries, an attempt being made to highlight those points likely to create difficulty (second part) before, in the conclusion, making certain proposals for action (2).

Generally speaking, the crossing of frontiers -- which makes it easier to cover up the tracks -- is a source of hightened difficulty. Company law also merits attention in this connection. Bringing subsidiaries and interests in other companies into play, clever set-ups may be arranged to facilitate deception, particularly beyond national frontiers. With generators, carriers, ancillary services and disposers working hand in glove, paperwork can be "fixed" and concealed operations organised by agreement among all concerned. But the purpose of this study is more limited. Whilst bearing in mind the implications of such factors as company and customs law on the solution of the problem of the transport of hazardous wste we shall confine our argument to the aspects connected with transport law.

2. The zero economic value of waste -- its consequences and the danger of fraudulent action

Apart from its possible recycling value, hazardous waste has no tangible economic value whereas many rules that are normally applicable to transport are explicable only in terms of the interest of the consignor, carrier or consignee in the satisfactory state of the cargo on arrival. In the present case this does not apply because no-one, except the disposer, has any interest in ensuring that the quantity of waste laid down in the disposal contract is dealt with and no economic agent has any personal interest in the fate of the cargo.

Thus, a major problem of the transport of hazardous waste is the removal of a temptation. It is tempting for a carrier, entrusted with conveying waste to a place of storage or treatment, to arrange for its disappearance en route. This saves him time, fuel and considerable effort, and enables him to underbid his competitors, a valuable advantage in a trade governed to a large extent by the market (3). Ways of achieving this end are numerous and often undetectable: in addition to unlawful dumping (4) he may dispose of the wastes in small quantities along a watercourse or a highway, using the drop by drop technique; he may also falsify the transport documents, either as regards the nature of the cargo (5) or concerning the toxic content of the cargo; lastly, he may dispose of the wastes by incineration (6).

This is the peculiarity of the problem posed by the carriage of wastes whether hazardous or otherwise. The carrier has no incentive to take care of it and sometimes may well benefit from disposing of it en route. Deliberate or negligent illegal action both lead to the same result: the hazardous wastes are disposed into the environment. The objective to be attained is therefore to ensure that the transport operation is carried through to a satisfactory conclusion, i.e. to provide for supervision and rules of liability such that the toxic waste entrusted to the carrier by the generator arrives in its totality at the agreed treatment or disposal site. (Subject to accidents or other unforeseeable and unavoidable events) (7).

3. Transport of poisonous wastes and transport of hazardous substances: obligations and liabilities

Poisonous wastes are hazardous substances the carriage of which is for this reason generally regulated by international or domestic law: the RID (Regulations concering the international carriage of dangerous goods by raid) as regards rail transport (8), the ADR (European Agreement concerning the International Carriage of dangerous goods by road) (9), the IMDG Code (International Maritime Dangerous Gods Code) (10) for international transport by sea, and the ICAO regulations (11) for international air transport.

However, these instruments, which do an excellent job of defining the precautions to be taken by the various agents involved in the transport of hazardous substances (12), will only be touched on in passing. The question put concerns not the obligations of the carrier of poisonous wastes, but liability resulting from failure to meet obligations and in general with the liability of all persons involved in transport whose default might create an environmental nuisance. In other words we are here dealing with "liabilty" rather than "responsibilty". Let us also remember that the relevant regulations take for granted that the carrier has an interest in looking after his cargo, which may not be so where the cargo in question is waste. Detailed examination of safety requirements prescribed by the various regulations will therefore be omitted.

4. Carriage of ordinary goods and carriage of hazardous wastes: contractual liability and damage to third parties

The specific nature of our problem from the standpoint of the carriage of ordinary goods also deserves to be streesed. Additional arrangements governing transport operations were devised from the standpoint of contract: they are essentially intended to regulate relations between parties to contracts involving transport. The relationship concerned are those of shippers, consignor and consignee with the carrier, and with other persons involved in transport. But the damage likely in the transport of hazardous waste is mainly by way of illegal acts and is caused to third parties not involved in th transport contract. The inclusion of damage caused to third parties among the risks of transport operations is basically a modern development at a time when the means of transport or the cargo has become capable of causing large-scale damage. The transport of oil by sea has in particular helped to promote this growing awareness. The traditional rules of transport law have thus gradually been extended to include new arrangements resulting from case law (e.g. the distinction between custody of a structure or custody of behaviour under French law) or from specific treaties (e.g. in regard to the transport of oil, nuclear or other dangerous substances) and intended to cover the new factor in transport constituted by the risk of pollution. This is thus our problem, a mixture of transport and liability law, namely to determine who, and to what extent, is inherently liable for the risk of pollution resulting from te carriage of hazardous wastes under the appropriate legislation, i.e. the person who, responsible for taking out insurance and supervising operations, will where necessary bear any penalties imposed and have to compensate any victims.

5. Criminal and civil liability - advantage of a mechanism allowing one person only to be designated liable vis-à-vis the victims

Liability of a person involved in a transport operation may be civil, with a view to providing compensation for damage suffered, or penal where an offence has been committed. It must be noted from the outset that the reasoning involved is very different in the two cases. In criminal cases individual liabilty can in principle only result from an offence committed by the person concerned. Criminal liabilty incurred in connection with the carriage of hazardous waste will thus vary in line with the obligations of those taking part in the transport oepration and sactioned by the criminal law: some dispersion is inevitable. On the other hand, in civil cases, it is common for the liabilty of one person to be borne by another, e.g. parents for children, principals for agents. In contract, th allocation of risks among parties to the contract often leads one to accept liability for all damage that may be caused by the other in the execution of the contract (e.g. the prime contractor for sub-contractors).

There may therefore be some concentration of liability on a single individual from the outset, thus facilitating legal action by victims. This kind of mechanism is of great value in the field of transport: the sequence of transport operations may involve a large number of persons performing physical functions (13) or legal (14) or commercial (15) functions or sometimes the last two together (16). Their liability, in line with te function they perform, will vary greatly in nature and extent (17). In addition it varies from country to country.

For victims, especially if they are outsiders unfamiliar with trade usage, the complex interrelations of those involved are incomprehensible and a rule channelling liability to one party to the transport operation will resolve uncertainty and indicate the person against whom action should be taken (18).

6. Insurance

The identification, beforehand, of the person responsible for damage occuring during carriage is particularly important because that person will have to have insurance cover.

In some Member countries, like Germany and Sweden, a carrier of hazardous waste has to produce evidence that he has special insurance. It is also a requirement elsewhere as it is for all carriers of hazardous waste.

The producer of the waste, too, has to have special insurance whenever the legislation of the country through which the waste is carried makes the generator liable for damage occuring during carriage (19).

Lastly, every economic agent concerned and likely to be held liable will also have to take out insurance, the risk represented by the carriage of dangerous waste rarely materialising but being extremely high (20). Apart from any direct damage that possible victims may suffer, the cost of cleaning up a contaminated area may prove considerable and in any case well beyond the financial capabilty of the person liable. So prudence alone should induce carriers and the like to be insured if the applicable legal system does not safeguard them from court action -- and a general review of applicable law in OECD Member countries seems to suggest that this is not generally the case for any of the economic agents involved in transport.

ANALYSIS OF LIABILITY INCURRED IN CONNECTION WITH THE CARRIAGE OF HAZARDOUS WASTE

While the transport operation is the most critical of the various stages leading up to the disposal of hazardous wastes, since it is the most difficult to control, breaking the load in the course of the operation may be a special cause of vulnerability. Only in exceptional cases is waste sent directly from its point of production to the point of disposal: it. is often carried by several means of transport in turn, e.g. rail, road, waterway. Between the various stages of the operation the waste is handled and stored (21). The risks of nuisance are then all the greater inasmuch as these accessory operations are outside the control of the carrier and depend on another transport agent who is not necessarily subject to such clearcut obligations. After analysing the liabilty incurred in connection with the transport operations itself (section 1) it is therefore necessary to take a special look at related storage and handling (section 2).

A. Liability implicit in the transport contract proper

i) Applicable provisions

Legal action against the carrier of wastes may be based on three types of legal provision:

a) Environmental protection law. This involves provisions relating specifically to wastes or to classified installations. In France, the law on classified installations (Act of 16th July 1976) does not apply to the transport operation itself since classified installations are necessarily immovable. Relevant provisions are to be found in the Act of 15th July 1975 on the disposal of wastes and the recovery of materials (22). Waste disposal here includes transport (23) and the Act requires transport undertakings to provide the authorities with full information concerning wastes accepted by them (Section 8). Undertakings are also in principle subject to certain special requirements (Section 9) (24) for certain types of waste.

In most of the OECD countries, special rules are to be found or are about to be adopted concerning the problems of hazardous wastes (25). As regards the member States of the European Economic Community they are moreover required to do so by the Council Directive on hazardous and poisonous wastes published on 20th March 1978 (backed up by specific directives of 16th June 1975 and 20th February 1978 on used oils and titanium dioxide wastes).

b) Transport law. In the absence of special provisions (26), liability for the carriage of hazardous wastes is a matter for the law dealing with transport by land (including inland waterway), air and sea. Unifying international agreements make provision for uniform solutions now in application in many OECD countries, but the systems that apply vary considerably from one mode of transport to another. In the case of combined or multimodal transport, where a succession of different modes is involved, it is therefore necessary to determine at precisely what stage in the transport operation the damaging event took place in order to decide who is liable and what liability system is applicable. Unfortunately, the ever wider use of containers makes this impossible. An international agreement designed to find a way out of this difficulty was adopted in Geneva in 1980 under UNCTAD auspices. Under that Convention, designed to provide start-to-finish rules in a single document for multi-modal transport when the point of departure or of delivery lies in one of the contracting countries, liability covering the whole of the transport operation is assigned to one and the same person.

Given the scarcity of provisions relating specifically to the carriage of hazardous waste, the Unidroit draft Convention deserves special mention. This 19-article Convention relates to civil liability for damage caused during carriage of dangerous goods by road, rail and inland navigation vessels. Going back to a 1972 Netherlands' initiative (27) the draft in its 1983 form seems to be largely based on the principles governing the 1969 Brussels Convention on civil liability for oil pollution damage (CLC) (28).

c) The ordinary law of liability (29). The French Act of 15th July 1975 states that its provisions "are without prejudice to the liability of any person for damage caused to others notably as a result of the disposal of wastes which such a person has transported" (Section 4/2). This merely reflects the general rule which will operate in all cases except where the channelling of liability allows for action against a specific person only [Cf quasi criminal liability) below].

ii) Liability: nature, extent and limits

There are three parties to a contract of carriage each of whole might, in the absence of special provision, be liable for damage occurring in the course of transport. The first is the carrier and the other two the consignor and consignee or, for our purposes, the waste generator and disposer, these two being treated as shippers for the purpose of the contract of carriage.

1. Where there are no specific provisions dealing with hazardous wastes

a) Party-to-party liability

Transport law generally imposes liabilty for faulty execution of the contract of carriage on the carrier, on whom lie certain obligations in terms of results with regard to the goods carried (30).

His legal responsibility is generally automatic for damage occurring in transit. The only difference here is in The Hague maritime rules under which the carrier, in principle liable, is merely required to show due diligence. This system, based on British practice, does not reappear in the Hamburg rules, due to take their place in the future. By and large, transport law does not require victims to establish that a wrongful act has been committed to be compensated (31).

However, a carrier may evade liability in certain cases which vary with mode of transport. For this he has, in some cases, to prove exonerating circumstances. These grounds for non-liability have to do with:

-- Events outside the control of the carrier (e.g. in The Hague rules, accidents at sea, aggression by public enemies, restraint of princes, strikes, riot, etc.).

-- The fault of the shipper (e.g. inadequate packing or marking, insufficient or erroneous descriptions of the goods).

-- The nature of the goods (including faults in the goods themselves) (32).

The air transport system and the Hamburg rules provide for a different system: the carrier gets the benefit if he can prove that he has taken every step that might reasonably be required to prevent the event and its consequences (33).

Lastly, the liability of the carrier is limited to amounts that depend on the nature, weight or volume of the goods carried, minimal ceilings being imposed by the law.

The shipper, as for him, is liable for the consequences of failing to fulfil his own obligations, provided the carrier so prove. Defective packing or marking (see article 12-4 of CIM, article 10 of CMR) or defective loading when this is the shipper's responsibility (e.g. carriage by rail wagon, voyage charter and inland waterway consignments, unless otherwise provided in the contract). Lastly, the shipper has to vouch to the carrier for the truth of his statements on the shipping document with regard to the quantity and nature of the goods and, in particular, their hazardous nature. Any false statement makes him liable to the carrier (see in particular Article 12 of the Convention on multi-modal transport) and sometimes even though the incorrect information may not have been the origin of the damage (see Article 31 of the French Act of 18th June 1966).

b) Quasi-criminal liability

The "damage" taken into account in goods transport law is essentially damage to the goods, items missing and lateness. They are confined to the relations between the carrier and the consignor or consignee. Injury caused to third parties during the performance of a transport operation generally comes under common law. A victim seeking compensation has to prove the liabilty of the causer of the damage -- the carrier, generator or other person. For this, the victim must, under the applicable national legislation,

216

prove that the author of the injury was at fault or had the safekeeping of the thing causing the injury (34). Some confusion in case law makes the outcome of such a case uncertain (35).

In some cases, however, the legislation makes provision with regard to hazardous transport, in order to organise compensation for injury to third parties. The liability of aircraft operators for damage caused on the surface is subject to special regulations in France (36).

For example, Article 12 of the Convention on multi-modal transport provides that the shipper share liability with the carrier towards third parties in the even of incorrect information on the transport document.

The increasing risks stemming from the international transport of dangerous substances has recently prompted lawmakers to make more systematic provision to facilitate the compensation of victims. Examples are the Conventions on the carriage of nuclear substances and, later, oil by sea (37).

A new Convention on the sea transport of dangerous substances other than oil is likely in the near future. In all these cases, the purpose is the same. To make it easier for the victim to take action, the law specifies the person liable for compensation even though he may have done no wrong (the operator of a nuclear installation or the owner of a ship carrying oil). The legislation lays down the conditions of and limits to his liabilty. If the action can be taken against the liable person defined in this way alone, to the exclusion of any other person, this is called the channelling of liability.

c) The question of channelling

The channelling of liability is a legal technique whereby the liability incurred on the occasion of an event is concentrated on one of the persons involved, thereby releasing from liability all other persons involved. The procedure is artificial to the extent that it means that a person is made liable for damage that has occurred neither through his fault nor his doing. But it is convenient, particularly since it simplifies the action to be taken by victims for whom there is only one liable person. This saves them, in cases of doubt, from having to aim their action in law in every possible direction. In addition it avoids the unnecessary multiplication of insurance policies on the part of potentially liable persons.

The acceptance of a risk by a person by whom it should not naturally be borne is not unusual in contract. Often, one of the parties will undertake to guarantee the other against certain forms of risk and make himself responsible for third party claims. This is a simple way of sharing the risk of a contract, a matter of negotiation between the parties. Any claim sent elsewhere than to the party bearing the risk concerned is immediately sent on to that party.

When this concentration of liabilty is imposed by law, the situation is completely different. For channelling (the term is confined to this particular situation) to operate, it is not sufficient for the law to identify the person on whom liabilty will be concentrated. Another -- complementary -- rule is required under which no action can be taken against anyone but the designated liable person. Victims, therefore, can take no action except in the direction laid down by the channelling law. This is relatively simple at the national level where the liabilty legislation can lay down that any other action, on any grounds whatsoever, is prohibited (38). In this system everything is settled. Internatinally, however, channelling is somewhat difficult to apply, for there is no way of preventing legal action on the part of victims not subject to the lawgiver. So to be able to function, channelling has to operate in a closed system, with all the countries that are party to the Convention that sets it up including a rule in their own legislation invalidating rules in conflict with that Convention. So far only the Paris and Vienna nuclear Conventions of 1960 and 1963 have such a mechanism. The 1969 Brussels Convention on liability for damage caused by oil transport fails to do so. It may concentrate liability on the owner of the ship but it does not prohibit all actions except those against the employees and representatives of that owner. Action can also be taken against:

a) all persons other than the agents or servants in countries that are party to the Convention, and

b) all persons in countries that are not party to the Convention.

Thus the 1984 revision of the 1969 Convention is designed to improve the mechanism by increasing the number of persons against whom action may not be taken. The purpose of the Unidroit draft Convention (see above) is the same Article 4, part 6 and 7 of the draft require that no claim shall be made:

1) against the carrier otherwise than in accordance with the Convention;

2) more generally, against:

a) the servants or agents of the carrier,

b) the pilot of the ship or any other person who performs services for the vehicle;

c) the owner, hirer, charterer, user, operator or manager of the vehicle, provided that he is not the carrier,

d) any person performing salvage operations with the consent of the owner, or

e) on instruction of a competent public authority,

f) any person taking preventive measures for damage,

g) any servants or agents of the persons mentioned under b) to f) above except in the case of inexcusable fault.

2.   Specific hazardous wastes legislation

Although, as we have seen, the disposal of dangerous wastes is covered by increasingly numerous special provisions in OECD Member countries, few of them would seem to specify liability precisely, particularly with regard to transport operations.

Most are content to spell out the obligations of generators, carriers and disposers without considering the consequences of failure to meet such obligations. In that case, responsibility naturally lies with the person wholly or partly failing to fulfil his obligations. Only where liability is specifically dealt with, do special rules replace ordinary transport law and modify the applicable liability regime.

Four solutions are theoretically possible, i.e. the generator bears all liability, the carrier bears all liability, the disposer bears all liability or they are all jointly and severally liable, each being responsible for all the damage caused. Where it does deal with the problem OECD legislation is inclined to concentrate liability on the generator alone, or make him jointly with the carrier. The EEC proposal for a new Article 15 makes the producer of hazardous waste responsible from the moment it arises until it is disposed of.

France has adopted a compromise solution, i.e. the generator is jointly liable with the carrier save where the former has used the services of an approved carrier who will then bear the entire liability (39).

Remark: One problem requires attention, i.e. both transport and hazardous wastes legislation contain mandatory provisions which are also binding on users. In the absence of rules specifying which body of law is to prevail, conflicts are likely to arise. An instructive example is to be found here in the maritime transport of nuclear substances. The adoption of the 1960 and 1963 nuclear conventions was without prejudice to maritime law, and since nuclear liability, which is much stricter than that for maritime transport, was inapplicable, serious difficulties arose (40).

iii) Sensitive issues

The variety of rules applicable, and particularly differences in grounds of exemption from liability for the carrier, stands in the way of any single clearcut solution in most cases. Among the difficulties with serious consequences for the carriage of hazardous wastes, five seem to call for uniform solutions:

1. The consequence of a fault committed by an employee who fails to perform his duties on the liability of the carrier who may or may not be required from case to case to provide compensation for damage caused by an employee acting without his knowledge. In France, the carrier has been held liable for pollution caused by the driver of a vehicle delivering fuel, who was in a hurry to dispose of his cargo (41), and for smuggling of which he had no knowledge (42).

2. The consequences of a fault committed by the contractor. Falso statements in the transport document are in principle the liabilty of the shipper (see above). However, proof may be difficult here: the distinction between mistake and deliberate deceit is a fine one: similarly the mistaken description of a substance may cause difficulty. The conditions in which the liability for damage would then cease to lie with the carrier would need to be spelled out.

3. Reservations entered by the carrier when he receives the cargo concerning the state of the waste, its packaging or labelling and the record of events occuring en route, may exonerate the carrier from liability on terms that vary considerably. There is here a regrettable source of uncertainty which a uniform drafting could have avoided. In general, standardization of the transport document seems desirable. (Computerization of the document should promote this.)

4. Limitation period: the respective liability of those involved once delivery has been effected could usefully be clarified, particularly where disposal is by dumping (43). In particular the period during which proceedings must be brought, and the way in which damage occuring after this period has expired is dealt with are problems resolved in a variety of ways by transport law or not resolved at all (cf. US Act on the dumping of chemicals).

5. The consequences of contracts for the hire of means of transport also need clarification, in particular the liability of the hirer in particular when he provides the driver.

B. Ancillary transport operations

Types of liability

The transport of hazardous waste, like any other transport operation, involves a number of persons with different functions and responsibilities. Subject to contractual provisions, which may determine liability in the absence of a mandatory statutory provision, three main categories may be distinguished for the purposes of liability (44). Some of these, such as warehousemen, and in general all those whose function is purely physical and precisely defined, are responsible for the consignment solely while it is in their hands and solely for some wrongful act or omission on their part (45). Others on the contrary have very wide-ranging duties. As for example a forwarding agent, responsible for arranging the entire operation: since he is free to choose routes and means of transport as he sees fit, he will be personally liable for acts of the carriers whom he employs (46) over and above his liability for his own acts.

A third category of person involved in transport operations, such as a commercial depository, is responsible for keeping and protecting the goods entrusted to him and is therefore personally liable for damage which occurs while the goods are in his possession. On the other hand, he will not be responsible for events occuring prior to or after such time (47).

Remark: Where the carrier, in addition to conveying the goods, handles and stores them in a place of transit he is required to comply with the provision applicable to such operations. Thus in France, his storage installation will be subject to the legislation on classified installations (48) or to the law on waste disposal. His position here will be no different to that of any other person performing the same function.

CONCLUSIONS: SOME PROPOSALS FOR ACTION

On 1st February 1984 the OECD Council adopted by Decision and Recommendation general principles applicable to transfrontier movements of hazardous wastes. The above analysis brings out a series of sensitive issues bearing on liability in regard to the transport of wastes, and concerning which it would be desirable for Member countries to adopt, if not a uniform position, at least converging views. These points are the following:

1.   Concentration of liability on one person: pros and cons

The dispersal of liability for damage that may arise during the transport of hazardous waste is a regrettable source of uncertainty:

a) for the victims who, to make sure they reach the person liable, have to undertake several lawsuits, and

b) for all the economic agents concerned who, under threat of having to pay out substantial compensationsimply by being closely or remotely involved in the waste disposal chain, have to have costly insurance cover.

It is arguable that someone responsible for the whole of the transport operation, from beginning to end, will be motivated to make the best choice of carrier and other services involved and to ensure that they perform their tasks satisfactorily. These are the reasons why recent Conventions on the international sea transport of nuclear substances and oil, with high third party risks, have endeavoured to adopt the principle of concentrated responsibility.

This mechanism would seem satisfactory provided it leaves room for the person specified as liable to take action against others involved who are guilty of negligence and provided it is backed up by a system of criminal liability whereby wrongful behaviour may be punished.

Its only justification is the special risk attached to the hazardous nature of the waste involved and therefore it would seem preferable that liability be only partly channelled, and solely for damage caused to third parties because of the towic nature of the wastes. The precedent represented by the carriage of nuclear materials points the way. In that case liability for radioactive injury is concentrated on the operators, that of the dispatching operator ending only or take over by the receiving operator. It is for this damage, therefore, and for this damage alone that the system of concentrated nuclear liability takes the place of the ordinary liability mechanisms, these remain fully effective for non-nuclear damage. Here, the generator or the disposer of the waste or even an intermediary given overall responsibility, and paid, for the transport operation could be treated as operator of the nuclear installation. Another point here is that this concentration of liability should apply not only to transport but to the whole of the "waste disposal chain" from waste generation to waste disposal.

However, channelling responsibility requires that the person liable be made objectively liable and in sufficiently widely-based conditions for victims to be able to expect reasonable compensation. In particular, cases of exoneration from liability must not be too numerous or too broadly defined nor must its limits be set too low. Otherwise the result will be exactly the reverse of what the system is intended to produce: since the victims can obtain compensation from it only on restricted terms and are unable, at the same time, to take action against any other person involved, the third-party risk in the transport of hazardous wastes is not properly covered. The system would then protect not the victim but the person liable where the situation would then be better than under common law (49).

2.   Harmonisation in principle with regards to the person or persons liable should be accompanied by harmonised solutions with regard to liability conditions and in particular with regard to:

i) the incidence of road accidents on liability and in particular external causes which in principle give exoneration from liability, e.g. force majeure, acts by third parties (theft, looting, sabotage, etc.);

ii) failure of a servant to perform his duties [above, see Sensitive issues].

3.   Transport is only one link in the waste disposal chain so it is clear that regulations govering it have to be co-ordinated with the other links. Of all the waste transported across frontiers, three-quarters is dumpted in the sea. A link-up between transport condition and sea dumping conditions (as set out in the 1972 Oslo and London Conventions) would therefore appear highly desirable.

4.   There would also seem to be considerable value in developing a simple, internationally harmonised transport document designed to receive the visa of each of the persons involved in turn, who would be required to enter any reservations they may have with regard to the state of the waste, its packaging, its quantity and soundness at the moment they take charge of it (severe penalties to be avoided for false statements).

5.   Lastly, particular attention should be paid to the reliability of staff, because experience shows that the satisfactory performance of the transport operation depends more on the skill and conscientiousness of carriers and their employees than on repeated detailed controls. At the moment, a special licence is required only in a few OECD Member countries (50). The EEC amended proposals for a regulation rightly makes provision for such licences (51). Both the carrier and handling agencies should be required to have such licences in every country at least for the transport of the most dangerous wastes. Special training with regard to the precautions to be taken, with particular reference to accidents, could be made a condition for being granted such licences. Conditions should be laid down for the withdrawal of the licence in the same way as for its issue.

1. There are many international agreements in which uniform solutions are provided, applicable to most forms of transfrontier goods transport, e.g. C.M.R. or 1965 Geneva Convention, for road transport, the C.M.I. or 1890 Berne Convention for rail transport, the 1929 Warsaw Convention for air transport and The Hague Rules or 1924 Brussels Convention and, for the future, the 1978 Hamburg Rules for sea transport. Multi-modal transport is covered by the 1980 Geneva Convention which has not yet entered into force. Most of these agreements have been amended on many occasions.

2. Transport, customs and company law. Where hazardous waste has to be transported across frontiers, it may be necessary prior to disposal, to obtain import, transit and export licences from the government concerned. For this reason, of course, the customs regulations for the operation of this licensing system has a direct impact on the obligations of the carrier (see report by J.P. Hannequart in this book).

3. See Hazardous Waste Legislation in OECD Countries, OECD, Paris, 1983 (p. 34).

4. See, for example, the case of halogenated hydrocarbons dumped in a watercourse in the Netherlands, or that of used oils dispersed on agricultural land in the United States. Document UNEP/WG 95/2, p. 7.

5. Even without making any false statement it is sufficient to present the product under an innocuous name: e.g. the term "paint sludge" used in the Raggi case heard by the High Court of Strasbourg on 11th March 1983 and by the Colmar Court of Appeal on 27th January 1984 does not reflect the highly poisonous nature of the waste in question (see Annex). On the description of poisonous wastes as "raw material", see UNEP document referred to in Note 4.

6. In 1975 the French authorities discovered that clandestine incineration of industrial wastes was taking place in the Mediterranean under cover of false bills of lading: this led to the Act of 7th July 1976 regulating the incineration of wastes at sea.

7. See report by J.M. Devos in this book.

8. The RID which constitutes Annex I to the International Convention for the Carriage of goods by rail was adopted in Berne in 1890.

9. The ADR, prepared in 1957 by a Working Group of the United Nations Economic Commission for Europe, came into force in 1968 (generally speaking international road transport are under the Convention on Contract for the international carriage of goods by road signed in Geneva on 19th May 1956).

10. The IMDG has been published by the International Maritime Organisation with a view to harmonising national legislation.

11. The technical instructions for the safety of the transport by air of dangerous goods (ICAO document 9284 AN/905) reproduce in substance Annex 18 to the Chicago Convention on International Civil Aviation adopted in 1981. (For an example of national implementation of the instructions see, in France, the Order of 14th January 1983, Official Gazette of 17th February 1983). These regulations replace the IATA (International Air Transport Association) provisions hitherto generally applied by airlines.

12. See report by A.I. Roberts in this book.

13. e.g. the handling firm.

14. e.g. the depository or the forwarding agent.

15. e.g. the forwarding agent concerned with transit.

16. e.g. the lighterage contractor in Mediterranean ports.

17. Cf. below, Ancillary transport operations.

18. This was precisely the aim of the reform of maritime transport of 18th June 1966: the carrier is liable for any damage occurring in the course of carriage, including loading and unloading operations, with the right to take action against other parties where appropriate.

19. See, inter alia, Belgian Act of 22nd July 1974, Art. 7/3.

20. See report by J. Deprimoz in this book.

21. See as an example of such risks, the case described in Annex I.

22. Among the decrees giving affect to this Act, the one of 19th August 1977 should be noted concerning information to be supplied regarding the wastes.

23. See Section 2/2: "The disposal of wastes includes collection, carriage, storage, sorting and treatment".

24. Since the decrees necessary to give effect to Section 9 have not yet been published, the provision is for the time being inapplicable (see in Annex, the Raggi case).

25. See in particular Section 4 of the Belgian Act of 22nd July 1974; the 1980 Canadian Act on the carriage of dangerous goods TDGA; the Danish Decree of 17th March 1976 on chemical wastes; The FRG Act of 7th June 1972 as amended on 21st June 1977; the Finnish Act of summer 1979 on waste management; the French Act of 15th July 1975; Japan; Luxembourg (Act of 26th June 1980); the Netherlands (Act of 11th February 1976); Norway (Act of 13th March 1981 not yet in force); Sweden (Order of 1976); the United Kingdom (Section 17 of the 1974 Pollution Control Act); and the United States. As regards the ADR and RID Conventions regulating the transport of dangerous substances, the replies to the OECD questionnaire seem to indicate that only the United States and Japan are not parties (no information on Canada).

26. As remains the case in numerous countries, particularly the developing countries. See UNEP/WG.95/4, page 3.

27. See preliminary study by M. EVANS, Secretary-General of Unidroit in Revue de Droit Uniforme, 1980, II, page 2 et seq.

28. See Unidroit Study L.V. - Doc. 46, Rome 1983.

29. See report by R. Lummert in this book.

30. For international transport (rules often included in national legislation) see: (road transport) article 17 of CMR, (rail transport) article 7 of CIM and (air transport) articles 18 and 19 of the Warsaw Convention of 12th October 1929.

31. The Hague rules also allow the carrier to be exonerated in the event of a fault of seamanship by the captain or crew.

32. V. CIM. art. 26s; CMR, art. 17s; Warsaw Conv., art. 18.

33. Article 5-1 of the Hamburg rules, see Article 20 of the Warsaw Convention. Although this mechanism seems to be less severe on the carrier it has sometimes been interpreted very strictly by the courts and may, in practice, end up by depriving the carrier of any grounds for exoneration whatsoever.

34. See report by R. Lummert in this book.

35. For example, French case law sometimes accepts the distinction between structural safekeeping and behavioural safekeeping. Structural safekeeping relates to a defect in the thing (the defect causing the injury) whereas behavioural safekeeping relates to the handling of the thing. French courts have used this distinction for accidents caused by objects that are inflammable or liable to explode. Its use would not seem possible for hazardous wastes because a special legal system (Section II of the Act of 11th July 1975) defines, in principle, the person liable (for general comments on this question see G. Viney, Traité de droit civil. Conditions de la responsabilté. Paris 1982, pages 800-806).

36. See Article L.141.2 of the Code de l'Aviation Civile.

37. The Brussels and Vienna Nuclear Conventions of 1960 and 1963 and the 1969 Convention on the transport of oil.

38. Section 42 of the French Act of 18th June 1966 on sea charter and transport.

39. See Article II of the Act of 15th July 1975 referred to above.

40. A protocol signed on 17th December 1971 resolved the difficulty by making the carriage by sea of nuclear substances subject to the nuclear conventions in regard to any radioactive damage.

41. Cass. Crim. 3rd 1979, Rev. trim. droit civil 1980, 117 dos. DURRY.

42. Cass. Crim. 18th June 1978, Rev. trim. droit civil 1980, 118. A change has occurred recently making it easier for the carrier to avoid liability. See Cass. Ass. plénière 17th June 1983, Rev. trim. droit civil 1983, 749, note by DURRY.

43. It would seem unreasonable to put then a liability on the waste generator who, having performed correctly his obligations, has lost all control on the waste discarded sometimes even a long time since. This difficulty is allowed for in the draft UNEP guidelines (UNEP WG 95/4). Para 45.3 of this draft provides that when the waste has been delivered in conformity with legal provisions to an agreed disposer, liability for damage caused by storing or disposing of the waste is borne by the disposer.

44. It is dangerous to attach too much significance to the terms designating each transport intermediary (forwarding agents, brokers or other) since this varies depending on modes of transport and usage. We are only concerned with the function and the liability attached to it.

45. The operator of a transit station where wastes are either simply stored or mixed with other substances intended to neutralise their effects or make them less harmful is only liable to his customer for some wrongful act or commission, save where the contract stipulates otherwise. So far as third parties are concerned, action may be brought against him under the ordinary law (private nuisance, nuisance, misuse of rights, custody of a noxious substance). In particular, he will be criminally liable for failure to comply with the law relating to classified installations and poisonous wastes.

46. See Article 99 of the French Commercial Code. To counterbalance this heavy responsibility he has a privileged position.

47. Liable in the same way as a carrier, a maritime depository may rely on the same grounds of exemption from liability as the ship owner vis-a-vis the plaintiff. (See Section 50 of the French Act of 18th June 1966).

48. Operation of which without a licence comes under Section 18 of the Act of 16th July 1976. Since a classified installation is necessarily mobile, even the means of transport will fall outside these provisions.

49. Because he would have the advantage of limits to his liability and in some cases, exoneration clauses that common law on liability would not permit. This is where the Unidroit draft Convention of 1983 seems to be at fault. Setting fairly flexible rules for the liability of the carrier, the text protects all other persons potentially involved in the transport chain from court action except in the case of inexcusable fault (see Art. 4, paragraphs 6 and 7 of the 1983 draft).

50. Including Germany, Japan, Sweden, see report by J.P. Hannequart in this book.

51. Commission proposal submitted to the Council of the European Communities on 24th June, 1983.

52.   The customs offences alleged, including false description of the material and a false statement of the actual consignee, were no longer punishable since the proceedings had been terminated by a compromise reached with the authorities.

Annex

THE RAGGI CASE

The "Société Alsacienne de Métallurgie" (SAM) imported from Germany 26 000 tonnes of toxic waste from 1977 to 1981 and dumped a large proportion of it illegally in a number of natural sites in Alsace. In a transit station operated illegally in Strasbourg the company received and stored, and mixed it with unburned residue before conveying it secretly to the dumping site. The only licence which the SAM possessed authorised it to store and crush metal residues (2nd class permit) but excluded any open air burning.

Raggi, the managing director of the SAM, was charged with seven offences before the Strasbourg Criminal Court, five of which concerned environmental protection (52).

On 11th March 1983 the court decided as follows:

The offences of:

1. Handing over waste or having waste handed over to a person other than the operator of an approved installation, and

2. Disposing of waste without the necessary approval under Sections 9 and 10 of the Act of 15th July 1975 could not have been committed, since the necessary implementing decrees had not been made.

But Raggi was found guilty of:

1. Operating classified installations without a permit;

2. Refusing to supply the authorities with information as to the nature, characteristics, quantities, origin, destination and disposal arrangements of the waste;

3. Contravening the provisions of the Order which in 1974 fixed the conditions for the grant of licences to treat metal residues. The penalty imposed was one year's imprisonment and a fine of 70 000 francs.

As regards the civil parties to the criminal proceedings, action by a Family Association was dismissed. On the other hand the merits of a claim by the French Federation of Nature Protection Associations were recognised and compensation of 50.000 francs awarded.

In addition the accused's driving licence was suspended and he was ordered to have the court's decision published in its entirety in a regional paper and in two leading national papers.

On appeal, the Court of Appeal of Colmar reversed the decision of the lower court on three points:

-- the decision to suspend the driving licence as an additional penalty was quashed (such penalty requires the use of a vehicle to commit the offence; this is not possible in relation to the refusal to supply the authorities with information, the only offence punished in this case).

-- while confirming the admissibility and merits of the claim by the French Federation of Nature Protection Societies, the compensation awarded was reduced to 15 000 francs.

-- the publication of the decision in the press was confined to the publication of extracts in a regional paper only and provided the cost did not exceed 5 000 francs.

LIABILITY OF THE GENERATOR AND THE DISPOSER OF HAZARDOUS WASTES

Rüdiger Lummert
Max Planck Institute für
Ausländisches und Internationales Privatrecht
Hamburg, Germany

SUMMARY

Liability for injury caused by hazardous wastes is found in different forms in OECD countries, either as liability for fault or strict liability. Liability for fault requires a breach of statutory duties or customary standards of care. Strict liability may be based on a general principle for hazardous activities or on individual statutory provisions. Liability of generators for fault does not necessarily end at the moment of transfer of the wastes to a transporter or disposer. If it is intended, however, to channel liability towards the generator, strict liability might be a more efficient instrument. Intervening unlawful acts of the disposer might relieve the generator of his liability, if they were unpredictable and irresistible.

INTRODUCTION

This report will examine the responsabilities of generators and disposers of hazardous wastes under the private law of OECD Member countries. All countries covered have their own constitutional, legal and administrative systems and legal traditions. They can be grouped into several "Legal families" such as the Romanistic, Germanistic, Anglo-American and Nordic family (1), but no common pattern exists a priori for a comparison of their rules on civil liability. The present report will undertake to develop a pattern for its limited purpose from the practical problems which occur in lawsuits and incidents involving hazardous wastes. A general discussion of the nature of legal obligations and the philosophy and principal features of each legal system will be avoided. Mention of such general issues will only be made where this is necessary to avoid misconceptions.

Hazardous waste has by definition a potential to cause injury to human health and damage to the environment. If such harm occurs, the question arises as to who will be required to pay the cost of restitution and cleanup.

According to the widely accepted "polluter pays principle" it would seem natural to impose this liability on the person who actually discharged the waste in the environment, i.e. the disposer. Such a simple solution would, however, ignore the complexity of the origin and management of hazardous wastes. Hazardous waste is not in itself a necessary good for society. Its generation should be avoided if possible. If it cannot be avoided, the generator should have a responsibility to render it harmless. This basic responsibility would not be altered if the generator uses the services of an employee or a contractor for disposal. Thus, if the generator is not himself the disposer, the distribution of responsibilities should take into account the preventive effect that civil liability may have on the production of hazardous waste.

National law may impose civil liability for a variety of reasons: a person may have caused harm to another through his own act or through a thing (an animal, an installation or otherwise) under his control. The liability for this harm may be based on fault (intent or negligence) or independent of fault (strict or absolute liability). Since the actors in hazardous waste management will mostly be companies, it is also necessary to determine who is responsible for acts of employees and contractors. In addition to the rules of private law, administrative regulation of hazardous wastes can be expected to play an important role in determining responsibilities for the prevention of harm.

Since all OECD Member countries are confronted with similar types of problems, it can be expected that their responses will share some common features. To the extent that equal or similar solutions were developed, it seems unnecessary to give a detailed country by country report on statutory provisions and legal construction. Such solutions will be described in general terms, while unique or unusual solutions by individual countries will receive particular attention.

LIABILITY FOR INTENTIONAL POLLUTION

In all legal systems, a claim for compensation is granted for personal injury and property damage intentionally caused by another. This rule applies also where the damage is caused by environmental pollution. To apply in these cases, however, the law would require evidence that the polluter has wilfully and knowingly inflicted an injury upon the victim. The mere intent to dispose of hazardous waste would not satisfy this requirement. Nevertheless, since virtually all OECD countries have prohibited the illicit disposal of hazardous wastes in the environment (2), the intentional violation of this rule can be an independently actionable offense. This liability would cover both personal injury and property damage. The question whether or not the cost of cleaning up the environment can be reimbursed will be discussed later.

1. Liability of the Generator

If the generator is not identical with the disposer, he is liable for intentional pollution only if he participated in the act of illegal disposal,

231

either physically or indirectly by giving the disposer directions or agreeing with him on using an illegal disposal site. In theory, there may be cases where the generator alone is liable, e.g. if the disposer does not know that he is handling hazardous waste, because the generator has deceived him.

2. Liability of the Disposer

A disposer who knowingly and wilfully discharges hazardous wastes in the environment without lawful authorization or other justification would be liable for damages. While the rule of law is simple in this respect, it may be difficult to prove that the act was intentional. If it cannot be shown that the disposer had knowledge of what he was discharging or that the disposal was deliberate, he could still be liable for negligence.

LIABILITY FOR ACCIDENTAL POLLUTION

If harm is caused by an accidental discharge of hazardous wastes or otherwise without the intent of the generator or the disposer, an action against them may lie if they are found to have violated a duty to take care. This liability for negligence is found in different forms in all legal systems (3). At Common Law, negligence is both an independent type of action and a standard of liability. In the Civil Codes of continental European countries and of Japan, liability for negligence has found a statutory base, which is supported by a large body of case law. In the Nordic countries, liability for "culpa" is based on customary law, supported by framework legislation and special statutory provisions. The general standard of care in all countries is the behaviour of a "reasonable man in the community". With respect to hazardous activities, many countries recognize an obligation of a person who exposes the public to a certain hazard to take all necessary measures to prevent or minimize this hazard. More specific obligations were developed in case law relating to various specific types of hazard. Provisions of administrative law aiming at the protection of man and the environment may be relevant in civil liability cases, since their violation may indicate the neglect of a duty to take care.

1. Liability of the Generator

The generation of hazardous waste is not in itself an unlawful act. It is normally the result of a legitimate economic activity. Nevertheless, in engaging in this activity, the generator creates a source of hazard and is therefore responsible for taking precautions against harmful effects. He may be required, according to administrative regulations or to the circumstances of the case, to recycle or reprocess hazardous wastes or to dispose of them in a safe and orderly manner.

In selecting the proper mode of disposal, he has to observe the relevant administrative provisions, the standard of customary practice and any particular indications of a specific risk. These standards may change rapidly in the light of experience gained. A generator who fails to take these developments into account would most probably be found negligent in all countries, if his hazardous wastes cause harm to other persons. On the other hand, if alternative methods of disposal are available which are considered equally safe, the generator has a choice according to his own priorities. If the disposal leads to unexpected harm, the mere fact that an alternative existed does not render his conduct negligent.

In complying with his obligations, the generator may use the services or a professional disposer or of a publicly owned disposal facility. In this case, the generator can, as a rule, not be held liable for negligent conduct on the part of the contractor. However, case law from several countries indicates that the generator's own duty to take care includes a requirement of careful selection and a necessary minimum of supervision of the disposer (4). Thus, if the generator knows or has reason to know that the disposer is unreliable or will dispose of the waste in an illegal manner, he may be held liable.

If the disposer causes harm to third parties because the generator failed to comply with his duties, the result is the same. If the generator fails to declare or label the hazardous waste correctly and the disposer fails to verify the contents of the consignment, both may be held liable.

The responsibility of the generator for supervising the disposer is a limited one. If the disposer runs an independent business, the generator cannot be expected to know about all its internal proceedings. Therefore, not every violation of the duties of the disposer can be held against the generator. Rather, a violation of his own duties has to be established. By this standard, a generator may be found negligent if he fails to verify if the disposer holds the necessary permits, if the premises of the disposer are unsuited for the intended disposal and the generator could have noticed this by a simple inspection, or if the generator selects a disposer with a poor reputation.

Specific provisions of administrative law which can be relevant to civil liability include:

-- a duty of the generator to transfer hazardous wastes to an authorized disposer, or to obtain written confirmation of acceptance by an authorized disposer before shipping a consignment of wastes (5) ;

-- a duty to declare the content of any consignment to the authorities, to label the consignment accordingly and to keep records of the generation and disposal of hazardous wastes (6).

Further duties of similar relevance may be created by national administration laws, as OECD Member countries go about implementing the Council Decision and Recommendation of 1st February, 1984 on Transfrontier Movements of Hazardous Wastes [C(83)180(Final)]. A generator who violates these duties may be held liable for any accident during or after disposal, even if the accident happens after the transfer to the disposer. However, this does not apply where there is no causal relationship between the specific violation of a legal duty and the accident. If hazardous wastes are leaked into the environment from the site of a disposer while they were preliminarily stored to await a permit for final destruction, the generator may be found negligent, because he did not ensure in advance that the permit was issued. It is more difficult to say if the generator can be held liable in case of disappearing of wastes, if he did not comply with recordkeeping requirements. The costs of a search for the lost consignment may be attributed to him, but the harm that the wastes may have caused to other persons would only be attributable if it could have been avoided with the help of a proper record.

2. Liability of the Disposer

Like the generator, the disposer is subjected to the standards of care under both civil and administrative law. He is responsible for the proper operation of his facility and for compliance with relevant statutory provisions. His responsibility may include precautions which concur with responsibilities of the generator. Thus, if the generator is liable to declare the composition of his hazardous wastes, the disposer may have an obligation to examine and verify it.

If the generator has complied with his duty to take care, and the disposer causes harm to a third party by a violation of his duties, the disposer alone is liable to pay compensation. Even if the generator did not satisfy all legal requirements, the conduct of the disposer may act as a superseding cause, which interrupts the chain of causation between the generator's conduct and the harm. However, not every misconduct of the disposer relieves the generator of his responsibility. The theories of causation which govern the imputability of civil liability vary widely from one country to another (7), but in practice, the following criteria are often used to distinguish between a contributing and a superseding cause: the events that interrupt causation can be described as (a) free, unconstrained and responsible acts of the injured party or third persons, (b) natural events, (c) an abnormal conduct of the injured party and (d) unusual coincidences (8). In application of these concepts, the generator of hazardous wastes would be free from liability if the entire harm would have been caused in the same way without his own negligence, and if the actual course of events was not foreseeable to him.

One example where a new chain of causation is started is the mixing of wastes by the disposer or collector. If the generator of a consignment of wastes had no knowledge that his wastes would be mixed with others, and the mixture has hazardous properties that were not present in any of the consignments that were mixed, the disposer or collector may himself be considered as a generator.

STRICT LIABILITY

The liability rules discussed so far were based on the assumption that a person can only be expected to bear the costs of harm inflicted upon another, if the action or omission which caused the harm was objectionable (liability for fault). In certain cases, however, liability can also be imposed on persons whose conduct was lawful but led to injury to another (strict liability). Typical cases where many countries have introduced strict liability are traffic accidents caused by motor vehicles and injuries caused by animals. In most countries, strict liability applies only where this is provided by a statute. Some countries recognize a legal rule of strict liability for abnormally dangerous activities (9). An express statutory provision for strict liability for hazardous wastes is found in only one country, Belgium (10). Several countries have such provisions in their environmental protection laws, especially in laws on water management or water pollution control (11).

In countries where strict liability has not yet been introduced or recognized, a tendendy can be observed towards the approximation of rules on liability for fault to those of strict liability. It has been argued that wherever a lawful but hazardous activity has led to injury or damage, the courts have found a duty to take care which the owner or operator of dangerous installation had violated. In several countries, the law on neighbourhood relations was used to establish rules, which make the payment of damages dependent on a balancing of interests rather than a showing of fault (12). In other countries, presumptions of fault (mostly rebuttable) and similar instruments were used to alleviate the burden on the victim.

1. Liability of the Generator

Under Art. 7, para. 3 of the Belgian law on toxic waste, the generator is responsible "for all damages whatsoever which could be caused by the toxic wastes, especially for the entire duration of their transport, by their destruction, neutralization or elimination, even if he does not carry out these activities himself". This provision is not only unambiguous in channeling all civil liability to the generator, it also introduces a relatively broad standard of causation (could be caused). If there is still room for superseding causes and liability of others for intentional torts or similar, the burden of demonstrating such particular circumstances would seem to fall on the generator.

The statutory provisions of other countries on strict liability may concern the generator as long as he is in possession of the wastes. E.g. he may be strictly liable for water pollution, if wastes are discharged into the water from his own premises, vessels or vehicles. After the transfer to a transporter or disposer, a claim against the generator under these provisions would be foreclosed. This applies to the air and water pollution control laws of several countries, but also to the French rules on liability for damage caused by things (13) and to statutory and customary rules on neighbourhood relations (14). At common law, and to some extent in the Scandinavian countries, strict liability is imposed on the owner or operator of an abnormally dangerous installation (15). This rule may apply to either the activity which produces the hazardous waste or to its disposal. In the first case, however, it is not sure if the off site disposal of the wastes can still be considered as part of the hazardous activity. In the latter case, the question arises whether orderly disposal in accordance with all relevant standards of care is necessarily an abnormally dangerous activity (16). Case law relating to these questions is not yet conclusive, but it seems improbable that strict liability will be widely used against generators after they have transferred their hazardous waste to an authorized disposer.

As was mentioned before, most existing provisions and rules on strict liability are applicable to the disposer rather than the generator. This is not the case under the Belgian law, but under most environmental protection statutes and under rules on liability for damage caused by things or on neighbourhood relations. Under Belgian law, the disposer can be liable by traditional standards (liability for fault), but this would not necessarily exclude liability of the generator. Under French law, if the disposer is considered the "guardian" of the wastes, this excludes the liability of the generator for the same cause. If the operation of a disposal facility causes inconveniences to the neighbourhood which exceed the ordinary measure of neighbourhood situations, this would form a separate cause of action. Under German law, the disposer would be liable independently of fault for equitable compensation, if his facility has a governmental authorization, but causes inconveniences to neighbours beyond the tolerable measure. At common law and under the legal principles of the Scandinavian countries, the disposer would be liable under similar conditions.

CAUSATION AND FAULT: BURDEN OF PROOF AND LEGAL DEFENCES

The generally accepted rule of law is that civil liability should fall on the person who caused by his fault an injury to another. As was mentioned before, various theories were developed to determine the concept of causation. Although these theories aim at explaining causation in scientific and logical terms, they all involve, more or less explicitly, an element of value judgment. From the complex causal relationships occurring in nature, they select the ones that should become legally relevant. So, not every condition contributing to an injury is considered a legal cause but only those that bear an adequate or substantial relationship with the resulting injury.

The assessment of causal relationships is particularly relevant for generators and disposers of hazardous wastes, if harm was caused by environmental pollution to which their contribution was relatively small or relatively remote. While the mere fact that several generators have contributed to a harm would not exempt any one of them from liability, the contribution of one of them might be so small that it becomes negligible in relationship to the harm ("one drum syndrome").

In some cases, causation of an injury is difficult to show, because the origin of hazardous wastes is unknown or because alternative possibilities of causation exist. In the case of alternative causation, each generator or disposer can be held liable, if his contribution could have caused the entire harm and his conduct gave rise to an action against him (hypothetical causality). In the case of wastes of unknown origin, the solution is more difficult. If wastes have disappeared from a generator or disposer, which could have caused the injury, statutory law may consider these as a legal cause. The wording of the Belgian law indicates that this is intended. In the absence of a statutory provision, the burden of proof would remain on the victim.

Proof of cause in fact can be facilitated by prima facie showings, "res ipsa loquitur" or similar. This means that the victim only has to show the first link in a causal chain that typically leads to the damage incurred. The presumption of causality is rebuttable, i.e., the defendant may show that in the actual case particular circumstances were present which interrupted the chain of causation. Similarly, prima facie showings may indicate that the defendant was at fault, which would reverse the burden of proof and require the defendant to exonerate himself by showing that he has complied with all standards of care.

In spite of the legal instruments already available, questions of causation and fault are still a serious obstacle to fair compensation of damage. With respect to the inherited problems of past hazardous waste disposal, this situation can hardly be altered. In the future, administrative records will be available in many countries to illuminate questions of fact. Where "cradle to grave control" of hazardous wastes was installed, evidentiary problems are likely to become less significant. Problems of proof of fault would become obsolete if more countries would introduce strict liability for these cases.

A limit to civil liability is generally drawn where the defendant has acted in good faith and could not have anticipated the course of events that led to the harm. Acts of God, natural events and acts of others including the victim may relieve him of his responsibility even in cases of strict liability (17). Acts of God and natural events typically include fire, flood, strike, riots, acts of war and sabotage. However, not every intervening act affects civil liability. For acts of God it has been pronounced that only such acts have this quality which are unpredictable and irresistable and originate outside the activity which is the cause of strict liability. Likewise, the behaviour of third parties and of the victim is irrelevant to the liability of the generator or the disposer of wastes, if he could have anticipated it or mitigated its harmful effect. Some statutes dealing with strict liability exclude acts of God and unintentional acts of third parties as legal defences. In particular, acts of one's own employees or contractors cannot be invoked as defences in many countries. In the case of claims under neighbourhood relation laws, it may also be a legal defence if the harm was not substantial with a view to ordinary local conditions, if the generator or disposer of the waste had used a suitable location for his purpose and carried out his activity in accordance with applicable rules and standards, and, in some countries, if he had obtained a governmental permit for his activity. The latter defence is available against claims for damages only in very few countries.

COMPLEX LIABILITIES

If harm is caused by acts of several persons, e.g., the generator of hazardous wastes and the disposer or several generators or disposers, two solutions are possible: the victim may claim compensation of the entire damage from one tortfeasor or a portion of the compensation from each of them. If the damage was caused by a conspiracy of several parties, e.g., if the generator and the disposer agreed to dispose of the wastes in an illegal manner, the parties of the conspiracy would be held jointly and severally liable, i.e., the victim could claim the entire compensation from any one of the tortfeasors, but he could claim the amount only once. If the tortfeasors participated in causing the same injury without being conspirators or joint tortfeasors, German law provides for joint and several liability by a statutory provision (18). At Common Law, joint and several liability would follow, if the injury to the victim was indivisible (19). In France, where joint and several liability is applied rather hesitantly in general, a statutory provision was introduced in the waste disposal law to impose such liability on generators and disposers, if wastes are handed to a disposer who is not properly authorized (20). In practice, joint and several liability was occasionally imposed by courts in environmental pollution cases, but if a reasonable basis could be found for an apportionment of responsibilities, this was the preferred solution, particularly in cases of property damage or economic loss (21).

If the victim's claims were satisfied by one of the joint tortfeasors, the question arises whether he can claim indemnification or contribution from other tortfeasors involved. The extent to which this is possible varies from one country to another. Where joint and several liability is easily available, it is the more important to distribute the burden fairly among the parties that are held liable. No uniform standards can be observed, but a certain preference seems to exist for giving a party strictly liable under a statutory provision a recourse for indemnification against another tortfeasor who was held liable for fault (22).

REMEDIES

The victim of an injury can claim recovery of his medical expenses, repair costs and similar costs of restitution, and, under certain conditions, also compensation of pure economic loss. Most countries have adopted statutory provisions for the recovery of cleanup costs (23). As a rule, these provisions require any person who caused environmental pollution by hazardous wastes to clean up the affected area, and authorize the competent agencies to effectuate the cleanup at the polluter's expense if he does not comply with this obligation.

As regards the responsible person, the laws usually put the obligation on the "polluter". In the case of pollution by hazardous wastes, it is not clear whether this means the generator or the disposer of the waste or anybody else responsible. One can only assume that in most cases the disposer is more likely to be held liable under such provisions, because he is immediately involved in the act of pollution. However, in some countries it is at least theoretically possible to hold the generator liable. In Germany, the obligation to clean up is subjected to the general rules of police law, which require "offenders" (Störer) to eliminate the hazards to public safety and order for which they are responsible. This rule may be enforced against the generator as well as the disposer, although the wording of the waste laws of the Länder seems to indicate that they are primarily directed towards disposers. In Belgium, the liability of the generator for cleanup costs is unaffected by the liability of others (Art. 7). In the United States, generators and disposers are both strictly liable for cleanup costs under sec. 107 CERCLA.

In Finland the person whose activity or proceeding has caused the pollution by waste is responsible for the cleanup. If the polluter cannot be identified, cleanup expenses may be imposed on the person responsible for the collection of wastes in the area (sec. 33).

In France, generators of waste can be held liable if they participated in a cover-up to withdraw themselves from the application of the law, e.g. through a sale or gift or other transfer of the waste (Art. 3). In the United Kingdom, the occupier of the land where wastes are deposited is responsible for cleanup unless he shows that he neither deposited nor caused or permitted the waste to be deposited. In any case, the responsible authority may recover the costs from the person who deposited or caused or permitted the waste to be deposited. In Denmark, the person who during storage, transport or disposal of chemical wastes causes pollution of ground water or soil, is responsible to follow the instructions given by municipal authorities for cleanup. If such orders are not complied within the time period specified therein, the authority may arrange for cleanup at the expense of the responsible person (sec. 4, para. 3 of the ordinance on chemical wastes in conjunction with sec. 49, para. 2 of the environmental protection act). In Norway, a similar rule was established by sec. 37 of the Pollution Control Act.

As a rule, the authorities can only enforce cleanup measures after having given the polluter an opportunity to take such measures himself. However, in some countries the powers delegated to the authorities are somewhat broader. In Belgium the power to organize the destruction of wastes is not conditioned on prior notice nor on an opportunity of the polluter to take action first. In Germany the authorities can proceed without prior notice if the ordinary way is likely to be overly expensive or unsuccessful (e.g. sec. 19 of the Bavarian Waste Law). In Norway the Pollution Control Authority can provide for cleanup measures without prior instruction to the polluter if such instructions would lead to a delay or if it is not clear who is responsible. In the United States, the President has the power to determine whether the government or a responsible party shall arrange for the cleanup (sec. 104, CERCLA). If he determines that a private party should take care of it, a certain period of time would be specified within which the measures must be taken. In Italy the public prosecutor can arrange for the cleanup at the expense of the responsible person, regardless of prior notice and an opportunity for private measures (Art. 9).

In certain countries not only government agencies but also private parties can recover their expenditure for cleanup measures, either under general rules of civil liability (24) or under special statutory provisions (25). In Norway polluters can be ordered by municipal authorities to pay reasonable costs incurred by another for cleanup (sec. 38). As a rule, such costs can be claimed if the responsible person has not acted within a reasonable time and the other party has taken measures to protect its property or to minimize or abate damage to that property, and if these measures have been executed with due care (sec. 76). In the United States, the government as well as private parties can claim recovery of their expenses from Superfund if they have acted in accordance with the national contingency plan (sec. 107(a) (4) (A) and (B), CERCLA).

239

Claims for recovery of cleanup costs are generally limited to "reasonable" amounts. While this is expressly stated in all legal provisions dealing with private claims, only the United Kingdom has an explicit provision giving the responsible person an objection against claims by the government for unnecessary costs. The rationale of the general practice seems to be that the responsible person is sufficiently protected by general rules of administrative law if the government takes action, while protection against excessive claims is needed in the case of a private party taking cleanup measures. Thus, the absence of specific language does not mean that the responsible person can be expected to bear cleanup costs beyond the reasonable.

Traditionally, it has been a problem of damage compensation law that environmental resources often cannot be attributed to private property rights and therefore are not compensable. Civil liability for cleanup costs is an important instrument for the protection of particularly those environmental interests which do not amount to private rights. Beyond this, an explicit provision for compensation of damages to environmental resources was introduced in the United States in sec. 107 (a) (4) (C) CERCLA.

In some cases of strict liability, the compensable amount is limited to a certain ceiling value. Such a limitation exists under international agreements on liability for nuclear installations and for oil pollution damage (26) and in some national laws. There are, however, examples to the contrary in national laws. The philosophy behind a ceiling value is that a person who can be held liable without his fault should be enabled to calculate his risk, especially in order to take out the necessary insurance. It appears doubtful, however, whether this argument is still in line with economic reality. Even if liability insurance for unlimited amounts is presently not available, this does not exclude the possibility of imposing unlimited liability -- even strict liability -- on generators of hazardous waste. The argument used in the context of nuclear installations and oil pollution, that the activities in question were necessary and beneficial for society and should not be discouraged, does not apply to hazardous waste which in itself is neither beneficial nor absolutely necessary.

In some countries, guarantee funds were created to ensure that financial obligations can be carried out even in case of insolvency of the responsible party (27). In the absence of a comprehensive fund system, compulsory insurance schemes might be used to secure payment in these cases. A fund or insurance scheme may cover governmental expenses for cleaning up the environment only, or it may extend to third party damage. The former approach was chosen by the United States, the latter by Belgium.

In order to prevent future damage, injunctions against the production of the disposal of hazardous wastes may be granted by courts along with the compensation of existing damages. While the generation of hazardous waste is not in itself unlawful, it may be restricted by administrative rulemaking and adjudication, or it may be ruled out by a court because it threatens to cause harm in a particular situation. With increasing use of powers by governments to restrict the generation of certain wastes, this option may become more significant in the future.

Transfrontier shipments of hazardous waste can lead to difficult problems in determining the applicable law and the proper forum in case of an injury. If a generator of hazardous waste in one country causes the waste to be illegally disposed of in another, it is up to conflicts rules to determine which law is applicable to a claim for damages. As we are dealing with torts, the conflicts law of most countries provides for application of the law of the place were the tort was committed (lex loci delicti commissi) (28). A different law may be applicable to claims arising from a contract between the generator and the disposer, but these claims will not be further considered in this paper.

The designation of the place where the tort was committed can be ambiguous if the responsible person has acted in one country and the damage has manifested itself in another. The various national laws provide different solutions to this problem. Some countries favour the law of the place where the damage occurred (e.g. France and Spain), others the place where the tortfeasor has acted (e.g. Austria and Italy), still others give the plaintiff a choice or apply ex officio the most favourable law (e.g. Germany and Switzerland) and some reject the entire concept of the lex loci delicti commissi and apply the law of the place of the court (lex fori) (e.g. the United Kingdom) or the law bearing the "most significant relationship" with the issue (United States) (29).

Many arguments have been exchanged about the pros and cons of each theory, and the debate in the litterature has to some extent influenced the interpretation of conflict rules. Therefore, the standards of conduct at the place of the act are considered even by courts who apply the lex fori or the law of the place of the damage. In some countries, a claim for damages will be dismissed if the act of the tortfeasor was justified by the law of the place of the action, others require similarity between the law of that place and the law which is applicable according to their conflicts rules (30). However, the practical results are still far from harmony. Given the differences between the substantive provisions, especially the attribution of liability to generators or disposers respectively, the question of the applicable law can be extremely important to the parties involved, and the results may be difficult to predict.

The selection of the proper forum can be of similar importance. Among the Member States of the European Communities, a Convention on jurisdiction and the enforcement of judgments in civil and commercial matters (31) provides that civil actions based on torts can be brought to the court of the place where the damage took place (Art. 5.3). This was interpreted to mean both the place of the act and the place of the result, giving the plaintiff a choice of forum (32). Judgements by a court chosen in compliance with this provision can be enforced in other Member States of the Convention. Outside the European Communities, however, it may not be easy to find the right forum, as some states protect their citizens against civil judgments by foreign courts, and a sentence favourable to the plaintiff may prove difficult to enforce.

The best response to these problems would be the international harmonization of legal provisions relating to claims for compensation of damages caused by hazardous wastes, including rules for the selection of the proper forum. Proposals for uniform rules were put forward in a set of Draft Guidelines for the Environmentally Sound Management of Hazardous Wastes by UNEP (33) and in a Draft Council Regulation on the Supervision and Control of Transfrontier Shipments of Hazardous Waste Within the Community, by the Commission of the European Communities (34). These proposals include provisions for strict liability for damages caused by hazardous waste (Art. 15 of the CEC proposal, Art. 45 of the UNEP proposal). In the related field of transportation of hazardous chemicals, international conventions have been proposed by the International Institute for the Unification of Private Law (UNIDROIT) (35) on behalf of the Economic Commission for Europe and by the International Maritime Organisation (36). However, in the case of the draft EEC Regulation which has meanwhile been transformed into a Council Directive, the question of strict and unlimited liability of the generator has caused much debate and could not yet be resolved. The UNEP draft is not yet ready for adoption, and the other drafts, although related in some aspects, deal with a slightly different issue, which implies a different view at the best possibility for channelling of liability (transporter vs. shipper).

CONCLUSIONS

Civil liability rules have essentially two functions: Compensation of harm that has already occurred and prevention of future harm. The findings of this report indicate that existing law on hazardous waste disposal is in many countries deficient in both respects.

For the victim of an injury, the most important issue is to get compensation and to be protected against future harm, independently of whether or not the generator and the disposer of a waste that caused the injury were at fault. However, in most countries liability is still based on fault. Moreover, the victim has to face the difficulty of multiple causation and apportionment of liability, i.e., if he has found a responsible party, he may only be able to recover a fraction of the damages.

For the full protection of victims it would be the most satisfactory solution to hold both generators and disposers -- jointly and severally -- liable for all damages regardlessly of fault. In this case the victim could be sure to get compensation if he can at least identify either the generator or the disposer. The question of final responsibility could be settled by way of indemnification or contribution after the claim of the victim has been satisfied.

From the point of view of a policy that aims at prevention of the generation of hazardous wastes, many existing laws channel the liability in the wrong direction. Rules on proximate cause as well as statutes on strict liability tend to favor the liability of the disposer over that of the generator. The rule of Belgian law which holds the generator liable is still a unique exception.

It is understandable that generators want to be spared from liability if they have complied with all relevant legal provisions regarding the handling of hazardous waste, and transferred it to an authorized disposer. It cannot be denied, however, that the inherent capacity of that waste to cause injury is produced, along with the waste, by the generator. Whenever actual damage is caused by such waste, its generator is accountable for at least a contribution to the damage. It would therefore not be against all justice to hold the generator strictly liable for this contribution. If a closer link exists between the damage and another contributing factor -- like negligence of the disposer or otherwise -- this can be considered in the context of indemnification. In the exceptional case that liability cannot justifiably be attributed to any involved party, financial guarantee schemes like a guarantee fund could be used to satisfy the victims' claims.

With respect to the internalisation of external effects, the economic efficiency of existing rules on damage compensation must be questioned. Financial arrangements for the recovery of cleanup costs can only work efficiently if they are supported by comprehensive liability of those who contribute to the problem. Channelling of liability to the generator and contribution by generators to a guarantee fund are possible economic tools for both prevention and cure.

NOTES AND REFERENCES

1. This classification was adopted from Zweigert and Kötz: An Introduction to Comparative Law. Vol. 1, New York, Oxford 1977. The author is aware that different systems for classification exist, most notably R. David: Major Legal Systems in the World Today. 2nd edition, London 1978, and the Volume II of the International Encyclopedia of Comparative Law, edited by the same author. The classification proposed by Zweigert and Kötz was followed here because it allows for more detailed differentiations.

It should be noted that at Common Law environmental pollution may give rise to an action for nuisance or trespass regardless of any consequential property damage or personal injury.

2. Statutory provisions against dumping of hazardous waste include:

Belgium: Art. 2, loi sur les déchets toxiques of 27th July, 1974 ;

Denmark: Sec. 3, ordinance on chemical waste of 17th March, 1976 ;

Finland: Sec. 32, law on waste disposal, as amended, 13th Feb., 1981 ;

France: Art. 2, Loi N° 75-633 relative à l'élimination des déchets ...
        of 15th July, 1975;

Germany: Sec. 2, waste disposal law, as amended, 5th January, 1977;

Italy: Art. 9, presidential decree N° 985, 10th Sept. 1982;

Japan: Sec. 5, para. 3, sec. 16, waste disposal and public cleansing
        law, N° 137, 25th December, 1970;

Netherlands: Art. 3, law on chemical wastes, 11th February, 1976;

Norway: Sec. 28 and 37, pollution control law, 13th March, 1981;

Sweden: (Prohibition of dumping of wastes in water only)
Law prohibiting dumping of wastes in water, 1971: 1154;

United Kingdom: Sec. 3, control of pollution act, 1974, ch. 40;

United States: (Prohibition of dumping under certain conditions),
Sec. 3008, resource conservation and recovery act,
21st October, 1976.

3. For details, see Limpens, Kruithof and Meinertzhagen-Limpens: Liability For One's Own Act. International Encyclopedia of Comparative Law, Vol. XI ch. 2. Tübingen, The Hague, Boston, London 1983. Relevant provisions and rules include:

France: Art. 1382 et seq. Code Civil;

Germany: § 823 Bürgerliches Gesetzbuch;

Italy: Art. 2055 et seq. Codice Civil;

Japan: Art. 709 et seq. Civil Code;

Netherlands: Art. 1403 Burgerlijk Wetboek

Sweden: 2 kap. 1 § Skadestandslagen;

Switzerland: Art. 41 Code of Obligations

Common Law Countries: Doctrines of Negligence, Trespass, Nuisance.

4. France: TA Versailles, 4th and 10th Sep. 1975, D 1976, 623.

Germany: BGH NJW 1976, 46.

Norway: Sunnhordland herredsrett, 30.10.1975, NRt 1976, 315.

United-States: Bianchini v. Humble Pipe Line Co., 480 F. 2nd 251 (5th Cir. 1973); City of Philadelphia v.Stepan Chemical Co., 544 F. Supp. 1135 (E.D.Penn. 1982).

5. Belgium: Art. 3 and 8, law on toxic wastes; Art. 16, enforcement
order of 9th February, 1976;

Denmark: Sec. 6, ordinance on chemical waste;

France: Art. 9 and 11, law on waste disposal ...;

Germany: Sec. 4, para. 3, waste disposal law;

Italy: Art. 10 and 16, presidential decree N° 985, 1982;

Japan: Sec. 12, para. 1, waste disposal ... law;

Netherlands: Art. 3, law on chemical wastes;

Norway: Sec. 28, para. 1 and 2, sec. 32, pollution control law;

Sweden: Sec. 14 ordinance on hazardous wastes, of 25th May, 1975;

Switzerland: Art. 30, para. 4, environmental protection act,
    Loi fédérale sur la protection de l'environnement,
    7th October, 1983;

United Kingdom: control of pollution (special waste)
    regulations 1980, S.I. N° 1709/1980;

United States: sec. 3002 RCRA and EPA Regulations, 40 CFR,
    para. 262.20;

European Communities: Art. 10, Directive 78/319/EEC.

6. Belgium: Art. 17 and 18, enforcement order 9th Feb., 1976;

Denmark: Sec. 5, para. 2, ordinance on chemical wastes;

Finland: Sec. 11a and 21, waste disposal law;

France: Art. 8, law on waste disposal etc., and enforcement
    decree N° 77-974;

Germany: Sec. 2 and 5, waste referral ordinance, 2nd June, 1978;

Italy: Art. 11 and 18, presidential decree N° 985, 1982;

Japan: Regulations pursuant to sec. 12, para. 2, waste disposal law;

Netherlands: Art. 4 and 5, law on chemical wastes;

Sweden: Sec. 8, ordinance on hazardous wastes;

Switzerland: proposed ordinance, not yet in force;

United Kingdom: special waste regulations 1980.

7. A.M. Honoré: Causation and Remoteness of Damage. In: International
Encyclopedia of Comparative Law, Vol. XI Ch. 7.

8. Honoré: loc. cit. (note 7) p. 47.

9. See the rule of the English case of Rylands v. Fletcher, L.R. 3 H.L. 330
(1868); United States: Restatement, Second, Torts § 524. A customary
rule of strict liability for dangerous activities applies in Norway
(Bedriftsansvar) and Sweden (rent strikt ansvar).

10. Art. 7 parag. 3 of the Law on Toxic Wastes of 22 July 1975 (Moniteur Belge
1975 p. 2365).

11. E.g.:

   Finland: chapter 2, § 2 of the Water Law;

   Germany: § 22 of the Water Management Law;

   Japan: Art. 25 of the Law on Prevention of Air Pollution, Art. 19 of the Law on Prevention of Water Pollution;

   Norway: § 9 of the Neighbourhood Relations Law;

   Sweden: § 30 of the Environmental Protection Law;

   United-States: Sec. 311 of the Clean Water Act; partly also Sec. 107 of the Superfund Act, which imposes strict liability for cleanup costs consistent with the national contingency plan.

12. Common Law countries: doctrine of nuisance;

   France: Troubles de voisinage, Art. 1382 et seq. CC;

   Germany: § 906 BGB;

   Norway: Law on Neighbourhood Relations;

   Sweden: Customary rules on neighbourhood relations (Grannelagsrätten).

13. Art. 1384 Code Civil.

14. See above, Note 12.

15. See references in Note 9.

16. This question was apparently denied in Ewell v. Petro Processors of Louisiana, Inc., 364 So. 2d 604 (La. Ct. App. 1978), cert. denied, 366 So. 2d 575 (La. 1979).

17. E.g. sec. 22, para. 2, German Water Supply Law; United States: sec. 107(b), CERCLA.

18. § 830 BGB.

19. Weir: Complex Liabilities, in International Encyclopedia of Comparative Law, Vol. XI ch. 12, p. 43.

20. Art. 11 of the Law on Waste Disposal, No. 75-633 of 15 July 1975, J.O. of 26 July 1975.

21. In the United Kingdom, apportionment was granted in Pride of Derby v. British Celanese, (1952), I All E.R. 1326 (C.A.), in a case of river pollution, while joint and several liability was imposed in a case of air pollution, Clarkson v. Modern Foundries, (1957) I.W.L.R. 1210 (Leeds Assizes). In the United States, joint and several liability was imposed in Michie v. Great Lakes Steel Division, 495 F. 2d 213 (6th Cir. 1974); cert. denied, 419 U.S. 997 (1974). The rule for

apportionment is contained in Restatement (2d) Torts § 433 A and B. In Germany, joint and several liability was denied in BGH NJW 1976, 797.

22. E.g. Germany: § 840 BGB; United-States: Adler's Quality Bakery, Inc. v. Gaseteria, Inc., 32 N.J. 55, 59 A. 2d, 97, 81 A.L.R. 2d, 1041.

23. Belgium: Art. 7, para. 2, Art. 15, 16, 18, and Art. 11 (guarantee
             funds), law on toxic wastes;

    Denmark: Sec. 4 and 49, para. 2, environmental protection law, sec. 4,
             para. 3, ordinance on chemical waste;

    Finland: Sec. 33, waste disposal law;

    France: Art. 3, law on waste disposal, etc.;

    Germany: Statutes of the Länder, e.g.:

Baden-Württemberg:	Sec. 10 waste disposal law;
Bavaria:	Sec. 12 and 19 waste disposal law;
Bremen:	Sec. 11 and 15 " " " ;
Hamburg:	Sec. 10 and 11 " " " ;
Hesse:	Sec. 10 and 11 " " " ;
Lower Saxony:	Sec. 4 " " " ;
North Rhine-Westphalia:	Sec. 15 " " " ;
Rhineland-Palatinate:	Sec. 11, 18 " " " ;
Saarland:	Sec. 5 " " " ;
Schleswig-Holstein:	Sec. 3

    Some of these include post closure cleanup; further authority
    is derived from the police powers of the Länder;

    Italy: Art. 9, presidential decree N° 985, 1982;

    Norway: Sec. 28, 37, 74 and 76, pollution control law;

    United Kingdom: Sec. 16, control of pollution act;

    United States: Sec. 107, CERCLA.

24. E.g., Germany: BGH NJW 1976, 46; France: TA Versailles 4 Sep. 1975, D 1976, 623.

25. Belgium: Art. 7, para. 2, Law on Toxic Wastes;

    United-States: Sec. 107 CERCLA

26. Art. 7 of the Convention on Third Party Liability in the Field of Nuclear Energy, Paris 29.7.1960; Art. V of the International Convention on Civil Liability for Oil Pollution Damage, Brussels 29.11.1969.

27. Belgium: Art. 11 of the Law on Toxic Wastes;

    United-States: CERCLA.

28. See e.g. Rest: International Protection of the Environment and Liability. Berlin: Erich Schmidt Verlag, 1978, p. 141.
Idem: The More Favourable Law Principle in Transfrontier Environmental Law. Berlin: Erich Schmidt Verlag, 1980. Ehrenzweig and Strömholm: Torts -- Introduction, in International Encyclopedia of Comparative Law, Vol. III, ch. 31. Tübingen/Alphen a.d. Rijn, 1980.

29. France: see Batiffol/Lagarde: Droit international privé. 7th edition, Paris 1983, part II, p. 246, N° 561;
Spain: see the text of Art. 10, N° 9 of the preliminary title of the civil code. The question is left undecided by Aguilar.
Navarro: Derecho civil internacional, Madrid 1975, p. 537.
Austria: Sec. 48, para. 1, Austrian conflicts law.

Italy: Art. 25, N° 2 of the preleggi al codice civile del 1942. The question is left undecided by Vitta; Diritto internazionale privato, Torino, 1975, Vol. III, p. 506. For further references see Rest: The More Favourable Law Principle ... (above N° 28) at 63.

Germany and Switzerland: customary law, see references by Rest, op. cit., at 59 and 61.

United Kingdom: Dicey and Morris: The Conflict of Laws. 10th edition, London, 1980, Vol. II, p. 927 et seq., rules 171 and 172.

United States: Restatement (2nd) of conflict of laws, para. 145.

30. British law requires that the act of the defendant is "actionable as a tort" also under British law; the United States rely on the similarity requirement. Art. 12 of the German introductory law to the civil code protects German defendants against foreign law which would impose obligations on them in excess of what German law requires.

31. Convention of Brussels, 27th September, 1968. Published in the Official Journal of the EEC, N° L 299 of 31st December, 1972.

32. Decision of the European Court of Justice of 30th November, 1976, Handelskwekerij Bier N.V. and Stichting Rijnwater v. Mines Domaniales de Potasse d'Alsace, Collection 1976, p. 1735.

33. Doc. UNEP/WG 95/4 of 15th December, 1983.

34. Draft Council Regulation on the Supervision and Control of Transfrontier Shipments of Hazardous Waste Within the Community, of 24th of June, 1983; adopted as Council Directive with some modifications on 28th June, 1984.

35. Draft Articles for a Convention on Civil Liability for Damage Caused During Carriage of Dangerous Goods by Road, Rail and Inland Navigation Vessels. International Institute for the Unification of Private Law (Unidroit), Rome, October 1983, Study LV Doc. 46.

36. See Sieg: Zum Entwurf eines Übereinkommens über die Haftung bei der Beförderung gefährlicher Stoffe auf See. Recht der Internationalen Wirtschaft 1984, p. 346.

LIABILITY OF TRANSACTORS FOR THE MOVEMENT OF HAZARDOUS WASTES

Jean-Marie Devos, European Council of
Chemical Manufacturers' Federations (ECCMF), Brussels

SUMMARY

The liability of transactors involved in operations for the movement of waste must be approached within the context of the ordinary law of liability and its application by the courts. Any international rules should be based on the fault or presumed fault seen as the occurence giving rise to liability. If "strict" liability rules were however to be adopted, they should be based on some logical attribution of liability. Liability must continue to be related to the fundamental concept of physical control over substances and surveillance of handling operations.

INTRODUCTION

The European Chemical Industry views symphathetically the efforts currently being made within a number of international bodies such as the EEC or, on a broader scale, the OECD, with a view to reaching a more satisfactory solution -- from both the technological and legal standpoints -- to a problem which has become particularly acute and has attracted considerable public attention in recent years. The improved protection of the human environment is indeed a legitimate concern of our industrial societies. In this respect, the question of the "management" and "treatment" of wastes deserves special attention.

The thinking under the guidance of OECD on the topic of transfrontier movements of waste is taking place at a particularly appropriate time in that it provides an excellent opportunity for an exchange and positive confrontation of views on the current situation and difficulties arisings. It should also make possible the much needed assessment of the value of present international initiatives which, it must be admitted, are tending to multiply without the necessary co-ordination always being present.

The considerations given below as a personal contribution are mainly devoted to issues related to the third party liability of transactors in matters of transfrontier movements of hazardous waste.

There are two traps which we must avoid falling into:

-- firstly, there is a real risk of <u>oversimplifying</u> the complex problems involved and the wide range <u>of situations</u> existing from country to country. Complexity is found on all levels -- technological, commercial, legal and administrative. This means that any attempt to establish international rules has to reflect this diversity and be somewhat flexible ;

-- secondly, there is a risk of exaggerating the <u>actual</u> hazards associated with movements of chemical wastes. While <u>all</u> industrial activity gives rise to waste, technological progress has nevertheless afforded increasing control over the negative results of production processes. Research and the development of industrial capacity to dispose of or neutralise hazardous wastes have made real progress and are encouraged both by industry in Europe and by governments. Finally, there is no reason why the transport of waste should be systematically considered by public opinion and by the authorities to be more hazardous than the transport of chemicals.

THE LIABILITY

It is nevertheless clear that the technological environment and everyday reality do influence measures taken at law on matters pertaining to transfrontier movements of hazardous waste. De lege feranda, the search for solutions at international level must form part of a continuity reflecting both the reality of the situation and the predominant legal systems.

In the light of comparative analyses of current national legislation presented at the Seminar on transfrontier movements of hazardous waste, it apears that -- fundamentally -- <u>there is no autonomous law of civil liability</u> applicable to the movement, treatment or disposal of hazardous wastes. The exceptions to the above observation are well-known and only a very small number of States have considered it necessary to adopt an entirely separate system of liability different to the ordinary law by imposing on an individual or more precisely on a single <u>transactor</u>, practically absolute liability even in the absence of any negligence on his part, or even for faults which he has not committed.

Some people have invoked the channelling of liability towards generators in regard to nuclear damage as a model for the problem before us. Such a system does not seem necessary nor justified in this case. Apart from its deliberately exceptional nature, the liability regime for nuclear damage applies within a particularly restricted context involving a very limited number of parties. Those parties are, in most cases, public or semi-public agencies and thus in the last resort liability is borne by the authorities or by the State as such. None of the above elements are present in the case of operations for the movement or disposal of waste.

Industrial reality outside the nuclear industry involves a vast number of individual transactions and a wide diversity of transactors. All this needs to be preserved, and does moreover reflect the basic principles underlying our market economies.

Industrial activity involving the movement of waste should, however, take place within a regulatory framework designed to <u>remove hazards</u> which it may entail for individuals and property and <u>prevent unlawful behaviour</u>. This has led the the European Chemical Industry to give its support in principle to efforts at international level, particularly within the European Communities, aimed at introducing arrangements for the surveillance and control of transfrontier shipments of hazardous waste.

There is no doubt that in economies which are more and more integrated one with another, international shipments of goods or other substances must be governed by rules which, if not similar, are at least compatible. Thus, although the chemical industry supported the underlying objectives and machinery in the Community proposal, it regretted the attempts to impose absolute liability on the waste generator.

In this respect, Article 15 of the Community proposal as laid out in its July 1983 version (1) was particularly unacceptable as a matter of principle and on legal grounds, but also for economic reasons. Article 15 provided for liability to be concentrated on the "waste producer", even though the framework and machinery of the proposal attribute distinct functions to different transactors. It is difficult to understand why in such a context one of the participants should assume liability for operations undertaken by other participants. Such a provision, too simplistic, did not and still does not have any part to play in an instrument unconcerned with the often complex questions of civil liability law.

The chemical industry expressly asked for Article 15, the implications of which had been insufficiently studied, to be withdrawn. Neither the need nor the legal basis for such a provision at Community level did not seem to have been proved (see Annex 1). It would be appropriate in this respect to examine whether national legal systems and existing insurance arrangements have really shown themselves incapable of providing adequate compensation for accidents occurring in the course of the transport and treatment of hazardous waste.

If some provision concerning liability were, however, considered necessary in an international instrument, such provision could only be envisaged with particular reference to legal principles common to the majority of States and to studies concerning liability currently under way in various international organisations.

The International Institute for the Unification of Private Law (Unidroit), which, with the support of the United Nations Economic Commission for Europe, is preparing a draft international convention on liability and compensation in regard to the carriage of dangerous goods or more precisely on liability in regard to the carriage of hazardous goods. Extremely detailed studies have been carried out in this field (2). Although the scope of the draft Convention is confined to civil liability for the carriage of dangerous goods, it is useful here to mention the main conclusions and working hypotheses adopted by Unidroit. The draft Convention is based on the principle of the liability of the person having the control and physical possession of the goods. In the transport context, the absolute liability of the carrier has been adopted without prejudice to any subsequent rights of action against third parties including the shipper and the manufacturer of goods. Provision is also made for compulsory insurance so as to protect victims against any risk of the person liable turning out to be insolvent.

The work of the Rome-based body could undoubtedly help the thinking of organisations which, like OECD and the CEE, have studied the issue.

*   *

*

Legislation and the courts in most of the States of the Community are generally very strict -- and rightly so -- where waste management and treatment are concerned. They do not however depart from the basic principles of civil liability, and any channelling of that liability is applied in a logical way. Indeed, even when fault is no longer the basis of liability, the latter continues to depend on the underlying concept to <u>the physical control of substances and surveillance of handling</u>.

This is a vital point. It is difficult to see how any legal system which is supposed to be fair and consistent can impose upon an individual or entity (i.e. the "generator") liability for the acts of <u>third parties</u> which cause damage. But this is nevertheless exactly what was proposed in Article 15 of the Community proposal. The nature of the liability envisaged was even more serious. It went beyond what is generally understood by "strict or absolute liability" since it took no account either of the person who actually caused the damage nor of the circumstances in which the damage had occurred and made no provision for any defence or subsequent proceedings by the generator against third parties.

In practice, the logic of the exclusive liability of "generators" would make the waste generator the sole actor in a one act play. It would be assumed that the generator alone had technical control and exercised surveillance over all operations leading up to the disposal of the waste.

Such an approach is not only at variance with reality but runs counter to all arrangements such as those envisaged in the OECD Decision or the Community proposal. Such arrangements are specifically intended to provide a regulatory framework for relations between a number of parties, primarily "waste generators", "carriers" and "treatment centres", subject to government surveillance. Although the parties concerned may in some cases be the same company responsible for one or all the operations, this is not generally the case. <u>The Community proposal moreover recognises this diversity and provides for prior compulsory approval of each of the parties involved to ensure that they have adequate capacity to carry through operations which, under their respective control, are intended to lead to the disposal or neutralisation of waste</u>.

It would, to say the least, be improper to try to make the generator liable for incidents occuring in the course of operations outside his control in cases where such generator had faithfully fulfilled his obligations under the regulations and called upon approved professionals. For example, it does seem normal that the carrier should bear the risks of the transport operation and any ensuing liability, since he has the physical, intellectual and legal control of the substances thanks, in particular, to the "consignment document" required under the Community proposal (obligatory information concerning the nature and characteristics of the waste and safety instructions). The same principle applies to operations undertaken by the "treatment centre" whose installations must be authorised. A different approach would run the risk of altering, to the detriment of safety, the behaviour of persons or firms who would no longer be responsible for any damage caused to individuals, to property or to the environment as a result of their own negligence.

In its initial version, Article 15 of the Community proposal did not recognise any of the grounds of exemption or possible defences usually available under liability law. Thus, neither "force majeure", "act of a third party", nor "fault of the victim" would remove or diminish the liability to be imposed on the generator. Is it adequately understood how discriminatory and exceptional a provision of this severity would be? This is all the more true to the extent that no ceiling or financial limit was specified even though such limits are necessarily and generally accepted under systems of strict liability.

As a result of pressure from the industry and on the request of several states and taking into account the comments from, the Commission, with the consent of the EEC Council, withdrew its proposal concerning liability. The Commission was instructed to submit more balanced proposals in this area within a period of four years from the adoption of the Directive (3).

CONCLUSION

As was previously indicated, one may question the necessity and foundation of any attempt to introduce a private law provision in an international instrument aiming at controlling transfrontier movements of hazardous waste. It may be doubted that public interest calls for such intervention; on the other hand, the same public interest calls for a satisfactory and coherent set of regulations on transport and disposal of hazardous waste.

In this respect it seems desirable that transport, treatment or disposal of hazardous waste be exclusively entrusted to private or public enterprises which have the necessary know-how and facilities to do so. Efforts undertaken at international level should thus essentially be directed towards preventive actions and developing administrative regulations.

If, however, for reasons of political opportunity, it were felt necessary to look for a specific liability system for movements of hazardous waste, the solution should not be one that departs from traditional law which is generally based on fault or presumed fault. Should nonetheless a system of strict liability be chosen, it would be important clearly to delineate the field of responsibility of each party involved in the operations affecting these movements. Clear determination is indeed essential as much for legal reasons as for economic reasons linked to technique and cost of insurance. In applying the principle of logical imputability, liability should be linked to the satisfactory implementation of the obligations which are imposed on each of the parties. Furthermore, such a provision should not prejudice the defendant to avail himself of the rights provided for his defense in most legal systems, such as act of a third party, victim's neglect or fault, act of God.

NOTES AND REFERENCES

1. Official Journal of the European Communities No. C 186, 12th July 1983 (Annex 2).

2. See amongst others document UNIDROIT 1983, Study LV, Doc. 46 containing draft articles for a convention.

3. This procedure is officialised in the new version of Article 15 as if taken up in the proposal for a Directive examined by the Council of the European Communities (Annex 3).

Annex 1

ECCMF PROPOSALS CONCERNING ARTICLE 15 OF THE EEC PROPOSAL

Article 15 (New)

1.　　The producer of waste, the carrier, the treatment centre or the licensed user within the meaning of (this directive) shall - as the case may be - be liable for the damage caused to a third party by this waste by reason of any violation of the obligation devolving upon each of them (respectively) under (this directive).

2.　　The agents and personnel of the producer, the carrier, the treatment centre or the licensed user are not considered as "third party" under paragraph one of this provision.

Comments

　　　It is important to define clearly the range of respective responsibilities of each of the parties involved in the operations described by the proposal.　Indeed, a clear definition is essential both for legal reasons and on economic grounds connected especially with the practicability and cost of insurance. Liability is therefore connected with the proper fulfilment of the obligations devolving upon the parties.　The proposal is therefore intended to impose upon each of the parties individual liability for the obligations incumbent upon them, any idea of joint liability being excluded, moreover, for the same reason.

Annex 2

TEXT OF THE EEC PROPOSAL OJEC, C186/9 (12th JULY 1983)

Article 15
(New Article)

　　　The producer of hazardous waste shall bear responsibility for the waste from the moment it arises until it is disposed of.

　　　He is liable for any damage caused to a third party by this waste.　The liability shall be independent from fault.

255

REVISED TEXT OF THE EEC PROPOSAL

(Directive of 6 December 1984)

Article 11

1.   Without prejudice to national provisions concerning civil liability, irrespective of the place in which the waste is disposed of, the producer of the waste shall take all necessary steps to dispose of or arrrange for the disposal of the waste so as to protect the quality of the environment in accordance with Directives 75/442/EEC and 78/319/EEC and with this Directive.

2.   Member States shall take all necessary steps to ensure that the obligations laid down in paragraph 1 are carried out.

3.   The Council, acting in accordance with the procedure referred to in Article 100 of the Treaty, shall determine, not later than 30 September 1988, the conditions for implementing the civil liability of the producer in the case of damage or that of any other person who may be to be accountable for the said damage and shall also determine a system of insurance.

THE CURRENT STATUS OF INSURANCE ON THE EUROPEAN LEVEL

John G. Cowell
Deputy Secretary General
Comité Européen des Assurance, Paris

SUMMARY

Possible changes in the rules governing liability and insurance raise important questions concerning scope of cover, limitation of liability and indemnity, cost, settlement of claims and enforcement of any compulsory insurance. Coverage of the transfrontier movement of hazardous waste must not only take account of the improved possibility of making claims but also the catastrophe potential which must be reflected in higher insurance costs. The present paper calls for the elimination of uncertainty, since, in insurance terms, uncertainty costs money.

FOREWARD

The present paper is based on ideas developed by European insurers on the insurance of liability arising from pollution and impairment to the environment. It is however the responsibility of the author alone and does not necessarily represent the views of the Comité Européen des Assurances or its member associations particularly in respect of the possible response of western European insurers to proposals for future action in the field of hazardous waste.

INTRODUCTION

Environmental impairment or, more particularly, the problem of hazardous waste is very much on the table for discussion at an international level. Indeed, it can be said that while the nineteen-seventies can apply be described as the "decade of defective products", the nineteen-eighties, under the influence of the "saga of the forty-one barrels from Seveso", look well set to become the "decade of hazardous waste".

The insurance industry in western Europe and in other parts of the world is ready to cooperate in seeking ways of dealing with the very real problems associated with the transfrontier movement of hazardous waste. The present paper focusses on the challenges to the insurance industry of possible changes in the rules governing civil liability in respect of the transfrontier movement of hazardous waste.

PRESENT SITUATION

The availability and cost of civil liability insurance, is essentially a function of demand. Hitherto, there has been no real demand for insurance of civil liability in respect of the transfrontier movement of hazardous waste. This is due to a number of reasons including:

-- the absence of compulsory insurance in respect of the movement of hazardous waste;

-- the limited number of undertakings involved in the transfrontier movement of hazardous waste;

-- the existence in all countries of western Europe of compulsory third party insurance in respect of the use of road vehicles with the courts showing themselves increasingly inclined towards an ever-wider interpretation of the link between injury or damage and use of the vehicle;

-- the lack of any apparent need for separate liability insurance cover.

The problem of the transfrontier movement of hazardous waste is not new. It has already been discussed in depth in a number of fora including:

-- the European Communities (in respect of the transfrontier shipment of dangerous waste);

-- the International Maritime Organisation (in respect of the carriage of dangerous goods by sea);

-- UNIDROIT (in respect of the carriage of dangerous goods by land) ...

The Comité Européen des Assurances has closely followed these discussions on behalf of insurers who feel that the case for compulsory insurance has not yet been proved. This does not mean that insurers are necessarily opposed to compulsory insurance but it does mean that they have an open mind on the subject. They would argue that it is not their task to suggest changes in the law but rather to comment on the likely economic effects of changes in the law on the availability and affordability of insurance.

Insurers have indicated to the European Communities and UNIDROIT that they would expect to be in a position to provide capacity and cover in line with the proposals currently being developed on liability and insurance. Capacity and cover could only be made available, however, subject to clarification on a number of key points concerning, for example:

-- channelling of liability;
-- heads of damage;
-- reasonable and insurable limits of indemnity;
-- cancellation.

CHANNELLING OF LIABILITY

a) Form of liability

Legal experts appear to favour some form of liability irrespective of fault.

Liability irrespective of fault is often referred to as "strict liability" (or "responsabilité objective" or "responsabilité causale"). This can be misleading since "strict liability" is essentially a court-made doctrine which can and does vary from one jurisdiction to another.

Liability irrespective of fault is not "no-fault" liability since fault of persons other than the person liable is often maintained, e.g. contributory negligence of the victim.

Indeed, fault of the victim or some other person is a frequent cause of loss which has led insurers to insist on the need:

-- to allow the courts to reduce compensation to victims who have contributed to their own loss;

-- to allow insurers to recover against third parties. Recovery may well have a beneficial effect on the cost of insurance (even though the benefit can only be measured over a relatively long period).

Above all, liability irrespective of the fault of a given person or persons should never lead to a weakening of the duty of care which is incumbent on all citizens in both a professional and private capacity.

b) Person liable

Liability irrespective of fault is intended to facilitate the payment of compensation to victims. The position of the victim is clearly made easier where he only has to address himself to a single person. Insurers have an open mind on whether liability should be channelled towards one person or another. Some insurers favour channelling to the generator (the waste producer) while others strongly favour the carrier. Intergovernmental organisations have adopted different views on this question. The European

Commission has suggested channelling liability to the producer while UNIDROIT (inspired by IMO proposals) has opted for liability of the carrier.

Carrier's liability has the attraction that it would complement existing liability under the road traffic laws while generator's liability has the advantage of placing liability on the person best able to support the cost of liability and insurance (or equivalent financial guarantee). What is important, however, from the point of view of both insurer and victim is the speeding up of the payment of compensation. Prolonged delays in the payment of compensation benefit neither party. One very obvious benefit of channelling liability is that it will reduce disputes about who is liable (1).

HEADS OF DAMAGE

Civil liability insurance is intended, amongst other things:

-- to guarantee the payment of third party compensation for injury or damage;

-- to permit the insured to transfer to a person or persons other than himself all or part of the economic risk (the non-commercial risk);

-- to allow the insured to avail himself of a wide range of legal and technical services offered by the insurer.

These considerations are of particular application to the transfrontier movement of hazardous waste.

Environmental impairment policies normally provide cover under the following heads:

a) Personal injury

Cover is provided in respect of bodily injury (death and temporary or permanent disablement) but may be extended to include mental injury and pain and suffering (pretium doloris) as well as pre-natal injury and genetic damage.

b) Economic damage:

Cover is provided in respect of direct property damage and direct damage to biotopes, flora and fauna as well as non-material economic damage including consequential loss.

The purpose of civil liability insurance is to pay third party losses for which the insured is liable under civil law. The starting point, however, is not whether or not civil liability insurance exists but whether the insured is (or should be) liable in certain circumstances. Above all, insurers look for clarity in determining their exposure to loss. A lack of clarity about what should or should not be covered can only lead to doubts and uncertainties and, in insurance terms, doubts and uncertainties can only mean additional costs. Expressions such as "any other loss or damage" or "prevention measures taken by any person to prevent or minimise damage or threat of damage" should be avoided.

i) "Any other loss or damage"

What concerns insurers here is the lack of precision coupled with difficulties of establishing causal link between event and damage where loss does not flow directly from bodily injury or property damage.

ii) "Prevention measures"

Liability insurers have become increasingly aware of the importance of loss prevention in recent years. Insurance is essentially intended to compensate those losses which, despite every effort on the part of the insured, prove unavoidable.

A distinction must be made however between measures taken by the insured and measures taken by any other person. A further distinction must also be made between measures taken before a loss and measures taken after a loss (salvage or clean-up costs). A liability policy is intended to compensate third party loss. It is not intended to compensate first party loss, that is, loss suffered by the insured. It is hardly reasonable to reduce the amount of compensation available to third parties by payments to the insured in respect of "preventive measures". Many insurers would, in fact, prefer to rate separately first party and third party costs.

Furthermore, many insurers would be extremely doubtful about covering under third party:

-- costs incurred by the insured before a loss where such costs formed part of the insured's normal duty of care;

-- costs incurred by third parties before a loss where the reality of a "threat of damage" may be difficult to determine.

REASONABLE AND INSURABLE LIMITS OF INDEMNITY

Insurance is essentially a mechanism for the spread of risk based on the principle of mutuality where the losses of the few are borne by the contributions of the many. The carriage of hazardous waste, unlike, for example, normal road traffic, presupposes the very real possibility of catastrophic loss resulting in the impairment of health and the environment over a wide area.

Normal road traffic accidents are characterised by two features: high frequency of accident and low severity of loss (amount). This facilitates claims forecasting. Hazardous waste losses are characterised, however, by low frequency of accident and high severity of loss. This can only complicate claims forecasting and make more important than ever the distinction between:

261

-- limits of legal liability and
-- limits of insurance cover (limits of indemnity).

There can be no question whatsoever of insurers providing unlimited insurance cover particularly in areas like environmental impairment where low frequency of loss makes it impossible to calculate effectively the insurer's maximum exposure. This, in turn, makes it impossible to obtain the global reinsurance support without which cover of potential catastrophe losses would quite simply be unavailable.

In the absence of compulsory insurance, insurance limits are normally freely negotiated between the insurer and the insurance buyer on the basis of:

-- individual estimates of exposure to loss;
-- availability and affordability of cover;
-- current levels of court awards ...

The above will further depend on such factors as:

-- type and size of undertaking;
-- type and size of waste load;
-- type of injury and damage covered;

-- identity and record of the insured ...

The above considerations would still have to be taken into account under any compulsory insurance scheme. The necessary distinction between legal liability (which might be unlimited) and insurance cover (which must be limited) can be maintained by the adoption of an appropriate wording, e.g. "the generator (or the carrier) shall be required to have and to maintain insurance or other financial security in an amount which not be less than ...".

Whether, in the final analysis, sufficient capacity can be made available will depend upon the answers to the above questions. Whether special arrangements, e.g. pooling and government guarantees, will also be required will depend upon individual insurance markets.

CANCELLATION

Cancellation of cover should always be possible in reasonable circumstances. Insurers would not wish to continue cover after receipt of notice of cancellation except in respect of carriage begun prior to receipt of notice.

Insurers would also not wish to find themselves obliged to accept risks they consider undesirable or to charge premiums they considered inappropriate to the risk. There is no reason, for example, by the poor risk (the risk with a poor claims record) should be able to profit at the expense of the good risk.

COSTS ASPECTS

What has been said so far is intended to emphasise the need for insurers to be in a position to calculate their exposure to loss -- particularly in the case of catastrophe exposures.

Insurance costs are governed not only by the number and size of claims made and paid but also by the need to investigate and defend claims which do not appear to be justified. Defence costs can and often do represent a major element in determining the cost of insurance. Once again this means avoiding unnecessary complications since complications can only be reflected in increased insurance cost.

WASTE PRODUCTS COVERED

Careful consideration will have to be given to waste products falling within the scope of any new rules on liability and insurance.

Insurers, while keeping an open mind on this question, would generally support the views of technical experts, e.g. within UNIDROIT, in favour of a list which is "neither too long nor too short" based, for example, on the UN Recommendations on the Transport of Dangerous Goods, 1983.

CLAIMS

As suggested earlier, the purpose of civil liability or third party insurance is to pay third party losses. Claims in respect of third party losses may be processed according to one of the following bases:

a) Act committed (fait générateur):

This basis is only used to any extent in Switzerland (except for some large contracts). What is important here is the cover in force at the time the act giving rise to the injury or damage was committed, e.g. faulty design.

In practice it may be very difficult to determine the precise moment of loss in the causal chain. In addition:

-- the insured may find that cover in force at the time the act giving rise to the loss was committed is quite inadequate in the light of monetary depreciation;

-- the insurer may find that he is liable for losses which are reported very late involving policies which expired some considerable time ago: in many countries this complicates claims reserving (2).

b) Losses occuring (survenance):

This is the most widely used basis in western European insurance markets. What is important in this case is the cover in force at the time the injury or damage occurred. Cover includes late claims which may go back over many years, e.g. Love Canal in the United-States.

Once again, it may be difficult to determine the precise moment of loss in cases involving hazardous waste, e.g. when the waste was first dumped, when the symptoms of injury first appeared or when the medical diagnosis was first made.

c) Claims made (réclamation):

Cover in this case is limited to claims which are made while the policy is in force. This basis facilitates underwriting since liability under the policy ends with the expiry of the contract although special arrangements can be made to continue cover for a limited period after expiry.

The insurer is, of course, liable for claims in respect of losses occuring before the inception of the policy but cover can sometimes be limited by fixing a retroactive date for the operation of cover. In the French market the claims made and losses occurring bases are frequently combined: the loss must occur and the claim must be made within the same policy period.

d) General comment:

The late settlement of claims is in the interest of neither the victim nor the insurer.

What is essential from the insurer's point of view is the ability to calculate his exposure and to reduce to a minimum the inevitable delays in settling claims which may be compounded by numerous factors including:

-- delays before occurrence of the injury or damage, e.g. unknown and unforeseeable side-effects developing over an extended period of time;

-- delays after the occurrence of injury or damage, e.g. identification of the person responsible, identification of the cause of loss ...

Insurers see advantages and disadvantages with all three bases described above. At present there is no consensus on which basis necessarily best protects the different interests of insurer, insured and victim.

SERIES CLAIMS

Multiple or series claims where, for example, any loss due to the same act or the same cause is considered to have occurred in the policy period in which the first loss occurred, are likely to be of importance in respect of the transfrontier movement of hazardous waste.

The expression "series claim" which does not necessarily figure "expressis verbis" in the standard policy conditions, is intended to facilitate the insurer's estimation of his exposure to loss. For the reinsurer, in fact, the series claim is a "catastrophe claim". Once again, both insurer and reinsurer must be in a position to limit their exposure. There can never be any question of open-ended cover ...

CRITERIA OF INSURABILITY

Environmental impairment including the transfrontier movement of hazardous waste raises important questions about the ongoing validity of traditional criteria of insurability which limit indemnity to losses which are:

-- sudden;
-- unintended;
-- unexpected.

Excluded, therefore, are losses other than sudden, unintended and unexpected losses, e.g. intentional act including intentional non-compliance with legal or administrative standards ...

Losses which are reasonably foreseeable cannot be covered. Losses which are foreseeable (and, therefore, avoidable) are no longer fortuitous and, as such, no longer insurable (3). Insurers insist on the ongoing importance of the absence of intention or expectation while recognising that the notion of suddenness may have to be re-defined to take account of losses associated with the transfrontier movement of hazardous waste.

CONCLUSIONS

Any change in the rules governing liability and insurance must necessarily take account not only of the availability and affordability of insurance but also its enforceability.

Careful consideration will have to be given to problems associated with the provision of cover with a high catastrophe potential at a price that can be carried by the insured undertaking;

-- with the enforcement of cover where, for example, the public authorities already have great difficulty in many countries in enforcing compulsory third party motor insurance.

Compulsory insurance inevitably raises the question of losses caused by uninsured or unidentified vehicles -- a question of particular importance in respect of road vehicles.

Provisions may have to be made to compensate such losses by means of a special fund administered by insurers.

Compulsory insurance of the transfrontier movement of hazardous waste (as proposed by the Commission of the European Communities) suggests a need for cover considerably in excess of the minimum amounts under the 1983 motor directive (4).

Cover must take account of the catastrophe potential associated with the transfrontier movement of hazardous waste which must be reflected in higher insurance costs since one catastrophe claim could more than absorb premiums accumulated over many years.

In the final analysis what insurers look for is clarity since it is the "grey areas" the areas of doubt and uncertainty -- which will inevitably complicate the administration of insurance linked to liability independent of fault.

Who should be liable? What level of insurance is required?

When should cover begin? When should it end? What happens where waste is loaded or unloaded by someone other than the insured? What happens in the case of piggy-back transport, e.g. on rivers or on the high seas?

These and many other questions will have to be considered before insurers can hope to respond on points of cover and cost. Uncertainty will always cost money. If we can never eliminate uncertainty (since the only certainty about the future is that it is uncertain) we must make every effort to reduce it to manageable proportions -- which is what insurance is all about.

NOTES AND REFERENCES

1. Disputes can never be wholly eliminated, especially in cases involving the movement and disposal of hazardous waste, shere pits, ponds, landfills, etc. may be used by a number of generators.

2. Such cases are usually known as IBNR (incurred but not reported) claims.

3. Exclusions common to all liability policies include:

   -- war, civil war, riot and civil commotion ...
   -- hurricane, whirlwind, flood, earthquake ...
   -- radioactive sources ...

4. The directive provides for the following minimum cover:

   -- in the case of personal injury, 350 000 ECU per victim;
   -- in the case of property damage, 100 000 ECU per claim.

   Member states may, however, set the following minimum amounts:

   -- 500 000 ECU in the case of personal injury involving several victims or
   -- 600 000 ECU in the case of personal injury and property damage ...

CONSIDERATION OF THE EXTENT OF INSURABLE RISKS
AND HOW THEY ARE TO BE COVERED

Jacques Deprimoz*
Paris, France

SUMMARY

Looking in this study at the carriage of hazardous waste by road only, it appears that in France cover for third party damage attributable to the load is provided mainly under motor insurance, compulsory since 1958, taken out by public or private road haulage contractors. In regard to consignments carried on vehicles requiring a "heavy goods" licence cover is unlimited in amount.

However, the consignor firm is not totally protected from subsequent legal proceedings and examination of insurance cover currently available to him reveals inadequacies and distortions which could be rectified by a few standard extensions to insurance cover without changing existing civil liability rules. It should be noted that the occurrence of damage which only becomes apparent long after the transport operation is completed does not seem to be as serious a problem as that arising in connection with final disposal.

In the increasingly frequent cases of transfrontier transport operations, French insurers will have to examine ways of providing start to finish cover by a single contract. In this respect, the international certificate of financial security recommended by the OECD Nuclear Energy Agency since 1968 for the transport of nuclear substances might be used as a model.

INTRODUCTION

Modern times, in which we have imperceptibly become accustomed to the concentration of risks in various forms, have already afforded several

---

* The author has been the Director of the French General Association of Accident Insurance Companies since 1971, of the French Nuclear Risks Insurance Pool since 1971 and of GARPOL since 1977.

examples of major disasters or near disasters. These have stimulated awareness of the need for regulation so as to prevent accidents occurring, prosecute offences and provide compensation for victims. France has had the Stalinon case, the Cinq-Sept fire, the Torrey Canyon and Amoco Cadiz oil spills and -- the most recent -- the strange adventure of the 41 barrels of dioxine from the ICMESA plant in Seveso which, for nearly a week in April 1983, spread -- if not panic -- at least confusion at the highest level concering their location in France.

It is true that several countries in western Europe affected by the transport of industrial waste by road already have regulations of varying degrees of strictness covering national transport operations. These regulations institute a system of notification of the competent authorities prior to the commencement of any transport operation; they enable a watch to be kept on such operations and ensure that waste does not disappear without trace. The relevant provision in France is Section 8 of the Act of 15th July 1975 and the implementing Decree of 19th August 1977.

However, the erratic route taken by the barrels of dioxine revealed the international nature of the problems arising. For technological and economic reasons firms generating waste often derive advantage from having their waste stored and neutralised at some facility abroad. Border controls are then seen to be highly uncertain. The consequences of all this are only too well known!

In the general interest, one can but welcome the good intentions of the Commission of the European Communities in tabling in June of 1983 a Proposal for a Regulation conferring on the 10 Member States of the Common Market certain "powers of action" in relation to the control of transfrontier shipments and the approval of disposers in the host country. One must also welcome those of the OECD Council in adopting in February 1984 a Decision and Recommendation on similar lines to be applied among the 23 Member countries of that Organisation.

*
*   *

What do insurers think of all this?

Whether a cause for indignation or satisfaction, the fact is that the Seveso dioxine has, not since April 1983, caused much concern to anyone other than insurance companies covering the civil liability of road carriers, consignors or disposers. No innocent person was injured in France. What is more, over a period of several years there has been no major disaster due to hazardous waste being transported by road (1), rail or waterway.

The Seminar organised by the OECD Environment Directorate as part of the work of the Waste Management Policy Group has at this very moment directed the attention of insurers, and especially French insurers, to a number of points:

-- firstly, some very useful quantitative information: in 1983, 5,000 tonnes of hazardous waste left France by road and 20,000 entered the country, i.e. an estimated 1,200 consignments per year in both directions by 20 tonne lorry. These data (2) are enough to show that the risk of accident already arises on a significant number of occasions;

-- secondly, the question of to which actor or actors in the "waste generator -- carrier -- disposer" chain they intend to give cover;

-- thirdly, questions concerning the limits to cover offered, per incident and over time under civil liability policies, and incidentally on the attitude of insurers as regards damage attributable to consignments of hazardous waste deliberately abondoned en route or diverted from their normal destination;

-- fourthly, since transfrontier shipments are the ones primarily concerned, the question whether insurers are prepared to cover without restriction accidents occurring outside the country where the insured person is established (generator or carrier).

*

*   *

Without prejudice to any official statement of their position by French insurers, the following lines include an initial descriptive section followed a second and third section which are speculative.

-- In Part 1, there is a general description of cover now available on the French insurance market for the reparation of damage caused to third parties and to the environment following a traffic accident involving toxic waste during transport within France.

-- In Part 2, also in connection with domestic transport operations, an outline is given of possible new types of cover designed in particular to make the waste generator aware of the hazards which he causes -- and perhaps of those caused by the carrier employed by him.

-- In Part 3, the special features of insurance regulations are brought out in cases where transfrontier shipments have to be covered from start to finish.

COVER CURRENTLY AVAILABLE

Having decided for the purposes of this study only to consider road transport, it is seen that cover for this type of risk concentrates primarily on the liability of the carrier within the framework of the motor insurance which has been obligatory since 1958.

This may be thought surprising. It is nevertheless directly in line with prevailing court decisions in regard to civil liability for risks created by the person who has custody of the object in question (Article 1384 (1) of the Civil Code): a person who agrees or decided to transport on a road vehicle on his own behalf or on behalf of another toxic products or waste is presumed to be liable for damage caused by such cargo. The vehicle and what it carried are so closely combined in regard to the causation of the damage that it seemed just to make the vehicle insurer responsible for payment of compensation in the first instance, in cases where courts apply Article 1384 (1).

However, limits to the duties of the vehicle insurer are carefully laid down by law:

i) Article R 211-11 of the Insurance Code deals specifically with damage "caused by the vehicle" (while moving or stationary on the public highway) in which connection inflammable, explosive, corrosive or combustive material is alleged to have provoked or aggravated the accident. It follows from this that damage attributable to the hazardous consignment, and which originates independently of the use of the vehicle is not covered by the vehicle insurance. This will be so in particular where damage occurs after the temporary or final abandonment of waste within or outside an authorised tip.

ii) A recent decree of 9th June 1983 extends the scope of car insurance to damage resulting from loading or unloading operations.

iii) Damage following the escape of products or waste should be broadly interpreted to include damage resulting from the dissemination in the air of toxic particles, travelling in some cases over large distances from the place of the accident.

iv) The amount of the cover deserves consideration. Generally, transport operations will involve vehicles the drivers of which require "heavy goods" licences (Categories C, D or E - Article R 124 of the Highway Code). In such case, vehicle insurance must cover all personal injury and material damage (including damage to the environment) without limitation of amount. Only in cases of vehicles which could be driven by holders of "light vehicles" licences would the cover be limited to Frs 5 million for personal injury and Frs 3 million for damage to property.

An attempt will be made below to measure possible damage.

This type of insurance, as is well known, is intended to cover the ordinary liability (tort or technical offence) of the carrier. All traditional defences are thus available (particularly in cases of contributory negligence by another driver in a collision) as are all subsequent proceedings for contribution (particularly against the loader where he has not packed the waste correctly or has not properly described the nature of the consignment).

It now has to be asked whether and how the waste generator, who is as a general rule the loader, can cover his civil liability on the French insurance market, whether this arises under a joint and several guarantee or as the result of some contributory proceedings. It has to be honestly admitted that, as things stand at present, cover provided by insurers for firms' civil liability may be less than totally watertight for at least three reasons:

1st reason: Many insurance contracts at present only cover damage attributable to activities undertaken on premises or construction sites of the person insured and, consequently, exclude "off-premises" risks in the course of carriage.

2nd reason: Where a company producing waste as a by-product takes out special insurance to cover its civil liability after delivery, such insurance will in principle only cover deliveries of marketable products the value of which is included in the figure for turnover used to calculate the premium. There is a strong likelihood that waste, the value of which is zero if not negative, will not, in the absence of some special provision, be treated as delivered goods.

3rd reason: Assuming that the two above restrictions can be removed by mutual agreement, it would still be necessary to provide express cover for pollution damage to extend even to cases where the cause was not an accident in the strict sense. Cover could be extended in this way to accidental damage to the environment of a prolonged or continuous nature during the transport of waste, a type of cover already provided in GARPOL-type contracts available on the market since 1977 for operators of installations classified under the Act of 19th July 1976.

TOWARDS POSSIBLE CHANGES IN INSURANCE
ARRANGEMENTS UNDER EXISTING LAW

The combination of carriers' civil liability insurance covering traffic accidents and waste generators' civil liability insurance covering any wrongful conduct in connection with the occurance of pollution damage during transport operations effected on their behalf presupposes that existing French civil liability rules will continue to operate whereby the determination of liability will depend on the circumstances of each individual case.

This status quo is moreover clearly maintained by Section 4 (2) of Act No. 75-633 of 15th July 1975 on waste disposal which stresses that the provisions of the Act "are without prejudice to the liability of any person for damage caused to another particularly as a result of the disposal of waste which has been in the possession of such person, has been transported by him or derives from products which he has manufactured".

The Act therefore rejects the concept of the concentration of liability at law on the waste generator alone, unlike the most recent Belgian provision (Section 7 (3) of the Act of 22nd July 1974 on toxic waste) and unlike the legislation of most European countries on the transport of radioactive products and waste, which adopts the so-called chanelling under Articles 4 and 6 of the Paris Convention of 29th July 1960.

Insurers unanimously take the view that, in the public interest, a monolithic solution should be avoided -- allowing for guarantee and contributory proceedings -- in relation to the reparation of damage occurring during or in connection with the carriage by road of hazardous products or waste.

From this standpoint, they see no reason to amend the existing law. Moreover, they point out that, in France, the law already adequately protects the interests of victims by:

-- either presuming liability due to a person having custody of the structure or of the behaviour of the dangerous object,

-- or involving the liability of the consignor for the bad choice of carrier to act on his behalf.

It is nevertheless true that the present insurance system could be improved by making the waste generator -- the principal -- more aware of his obligations and of the risks involved and, in particular, by getting him to subscribe to an amendment of his "operator's civil liability" policy. This amendment

-- would not only cover the liability of the loader throughout the operations, from leaving the factory to arrival at the premises of an approved disposer (including storage and handling time for transfers of load in multi-modal transport operations).

-- but would also cover, where necessary, the liability of the carrier designated and instructed by him, for damage caused or aggravated by the waste transported (which would of course leave traditional vehicle insurance to cover all traffic risks totally independent of the cargo and attributable solely to the vehicle).

The second element of the cover accorded by the amendment would, to the extent that it would be subscribed "on behalf of the carrier" be deemed to meet the obligation to take out vehicle insurance under Article R 211-11 of the Insurance Code. Such cover would be very similar to that recently made available on the French market for damage caused or aggravated by radioisotopes intended for industrial or medial purposes and transported by road (3).

In regard to the cover briefly outlined above, three questions seem to be of particular interest to the participants at the Seminar organised by the OECD:

-- What would be the attitude of the insurer in the event of deliberate abandonment of waste en route or after being diverted from its normal destination?

-- What would be the time limit for the insurer's liability for damage discovered after completion of the transport operation?

-- Will the maximum financial commitment be satisfactory?

Brief replies can already be given to these three questions:

i) Deliberate abandonment or diversion:

It has already been noted that the "negative" economic value of waste to be disposed of and neutralised may encourage this kind of improper practice. On this topic, the insurers' reply is bound to be a qualified one:

-- If such conduct is the willful and deliverate act of the waste generator or the carrier, it constitutes a fraud or deliberate fault the financial consequences of which cannot be borne by the insurer, on grounds of public policy;

-- If, on the other hand, the act is committed by employees of the generator and/or the carrier without their knowledge, insurance cover will still be available for victims.

What is more, if the suggestion made at the Seminar by Mr. Henri SMETS were to be adopted, advocating a financial penalty equivalent to 300 dollars per tonne payable by the carrier to the Treasury or to some special environmental protection fund in cases where waste fails to reach its destination and is not recovered within a short period, it is clear that the insurer could not accept responsibility for such fine.

ii) Differed damage

Everyone is aware of the difficulties, not to say the impossibility, facing insurers in covering the direct but distant consequences of the activity of policy holders when such damage occurs more than 5 or 10 years after the activity itself has terminated. This problem is presumably a major one in connection with the management of tips, deposits or silos for the permanent burial of toxic waste.

On the other hand, the need for extensive subsequent cover is not quite so obvious in connection with damage associated with consignments of waste to the extent that specific operations are concerned and that waste liable to harm third parties and the environment can be identified at all times thanks to the "consignment document" and the procedure for the return of this document to the consignor duly signed by the approved recipient (trip ticket system).

iii) The probable scale of accidents occuring during the carriage of waste

It would no doubt be rash to give over-precise figures.

To take the example of the carriage of radio-active substances, memories will go back to the emergency stop on the road from Marcoule to Malvesi in February 1975 which caused the ejection of two barrels of sodium uranate; decontamination and resurfacing of 400 metres of highway cost about 150,000 francs. There was also the collapse of a crane in a port while unloading magnesium uranate which involved about 200,000 francs damage to installations and neutralisation costs.

It is true that these reassuring examples concern products with very low radioactivity. Damage is likely to be very much greater with highly toxic waste. Present observations suggest three levels of potential accidents classified according to the medium of transmission of the toxic products:

-- If the medium is a watercourse, fishing losses may well be in the range of 100,000 to 1 million.

-- If the medium is a spring or ground water used for drinking water, the cost of pollution control and of direct losses to users could reach from 5 to 10 million francs.

-- If the medium of transmission is the air (a Seveso type accident) emergency measures and damage and injury to health and property could easily exceed 100 million francs.

On the existing French insurance market, GARPOL type contracts could easily provide cover of up to 30 million and perhaps even 50 million francs.

This seems to indicate the limits to the scale of risks which markets could finance without too much trouble at the appropriate time.

SPECIAL FEATURES OF INSURANCE OF TRANSFRONTIER SHIPMENTS

Can the carriage of waste across one or more frontier involving potential liability on the part of the generator or the carrier be covered by insurance from start to finish?

Such cover seems all the more desirable inasmuch as victims with a direct right of action against the insurer ought not to find themselves referred from one insurer to another depending on the exact spot where the damage in question originated. It is easy to imagine the difficulties that might arise in cases of the leakage of volatile substances onto both sides of a frontier?

But, to achieve this unity of cover "from loading to unloading" (which does not go as far as the term "cradle to grave" frequently used in relation to waste in the course of the Seminar) it is necessary to remove any obstacles in insurers settling claims abroad.

This obstacle has already been overcome in the case of civil liability insurance taken out by nuclear operators dispatching consignments of new or irradiated fuel, or even radioactive waste from one country to another. The insurance of the consignor remains effective up until the operator of the recipient installation takes possession, it being understood that the agency providing the insurance on behalf of the firm established in its own country will contract with a foreign insurer for the latter to settle on its behalf accidents occuring in the territory of such second insurer and will indemnify it in respect of settlements made (4).

It has thus become possible for the "Civil Liability" insurer of a firm which dispatches nuclear substances (whether or not it is itself the "producer" of those substances) to provide the firm with an international insurance certificate from start to finish of the operation in accordance with the model recommended since 1968 by the Steering Committee of the OECD Nuclear Energy Agency (5). Article 4 (c) of the Paris Convention of 29th July 1960 stipulates that this document must be handed over to the carrier and, more specifically, to the driver of the vehicle for the purposes of border controls.

Subject to drafting changes necessary to cover the transport of hazardous waste, a similar certificate could be made the standard document to accompany the trip ticket.

Requirements imposed in connection with the issue and inspection of this international certificate need not necessarily be seen as excessive if the number of cases to which they apply is borne in mind. In 1982 there were some 1,200 border crossings in and out of France by lorries transporting hazardous waste, while the statistics of the French Nuclear Risks Insurance Pool recorded 245 international shipments of nuclear substances (160 into and 85 out of France) for 1983. The figures do not differ all that much.

*
* *

What provisional conclusions can be reached?

In the light of the Decisions and Recommendations adopted by the OECD Council in February 1984, insurers are coming to realise the importance of the transport and disposal of inherently hazardous industrial waste. They are in no way seeking to avoid their financial responsibilties in regard to accidental damage attributable to these operations.

On what terms, within what limits? The search for the most appropriate arrangements needs further consideration and probably also consultation.

As at present insurers would simply like to say:

-- that it does not seem necessary to them, in regard to (national or international) shipments of waste, drastically to alter legal liability rules applicable to the parties to this type of operation,

-- that they could not be expected to intervene in cases of manifestly unlawful or even fraudulent conduct such as the deliberate abandonment of waste on an unauthorised site,

-- that the introduction of a new form of compulsory insurance does not to them seem necessary at either national or international level, and

-- that in any event they have an open mind towards any forward steps in the definition and forms of the cover they offer.

This study therefore has no ambition other than to indicate one of the possible ways ahead.

NOTES AND REFERENCES

1. Annex 1 contains statistics for 1982 for road traffic accidents involving hazardous waste, prepared by the CITMD (Extract from Transport Bulletin N° 2100 of 2nd May 1984).

2. Personal Communication from H. Yakowitz, OECD.

3. Annex 2 contains the model clause that the French Nuclear Risks Insurance Pool advocated on 23rd May 1984.

4. cf. JACQUES DEPRIMOZ: "Comment faciliter l'assurance Responsabilité Civile pour les transporteurs internationaux de substances nucléaires", R.G.A.T., October/December 1975 (pages 489 to 506).

5. Annex 3 contains the model ENEA Certificate in the two languages.

Annex 1

SOME DATA ON ROAD TRAFFIC ACCIDENTS INVOLVING HAZARDOUS SUBSTANCES

(from Bulletin des Transports No. 2100 of 2nd May 1984)

Statistics for 1982 established by the Interministerial Commission for the Transport of Hazardous Materials (35/37 rue Frémicourt -- 75015 Paris)

1. Classification of accidents by presumed cause

Causes attributable to the hazardous substances.	3
Causes attributable to the vehicle transporting the hazardous substances.	
-- Human causes	94
-- Equipment-related causes (broken coupling, burst tyre etc.,)	22
Third party negligence	74
External causes (bad weather, black ice, broken-down verge of highway etc.)	20
Unknown causes	20
TOTAL	233

2.   Spills, losses of load and leaks

Accidents characterised either by injuries attributable to the substances transported (burns, poisoning) or by a spill or loss of load -- excluding cases where the dangerous substance has remained neutral

-- Explosive substances	1
-- Compressed, liquefied or dissolved gas	4
-- Pyrophoric and liquid inflammable substances	57
-- Poisonous, corrosive substances etc.	20
TOTAL	82

Annex 2

CIVIL LIABILITY INSURANCE FOR HOLDERS AND USERS OF SOURCES OF IONIZING RADIATION

(outside a nuclear installation)

Model Clause

"Extension of Civil Liability Road Transport Cover for Sources of Ionizing Radiation"*

(To be included in the Special Conditions or by amendment to the General Conditions)

1. To comply with the obligation to take out insurance under the first sub-paragraph of Article R 211-11, of the Insurance Code, as amended by Decree No. 83/482 of 9th June 1983, the subscriber declares that the apparatus or containers containing the sources of ionizing radiation referred to in paragraph .... of the Special Conditions, may be transported on motor vehicles belonging to him:

-- between his own premises and those of the suppliers of the said sources situation, save where expressly otherwise provided, in metropolitan France or the Principality of Monaco,

-- between the places in which the said sources must be used situated, save where expressly otherwise provided, in metropolitan France or the Principality of Monaco.

He declares that these transport operations are effected in accordance with the Order of 24th June 1974 and its annexes.

2. In application of Article 5 of the General Conditions, it is expressly agreed that cover under the present contract shall be extended to the monetary consequences of any civil liability incurred by the insured for the damage defined in Article 1 of the said General Conditions, caused or aggravated by the sources transported.

This insurance shall be deemed to include, notwithstanding any contrary provisions, cover at least equivalent to that required under the Insurance Code for complusory motor vehicle insurance.

(*) Text recommended by the Pool français d'Assurance des Risque Atomiques on 23rd May 1984.

However, there shall be no cover under these provisions if, at any time of the accident, the vehicle transporting the source of ionizing radiation is not covered, for damage other than that for which reparation is provided above, by valid motor vehicle insurance.

3.   The present extension of the cover shall be granted in consideration of the payment of a premium to be calculated as laid down in Article ....

Annex 3

CERTIFICAT DE GARANTIE FINANCIERE

POUR LE TRANSPORT

DE SUBSTANCES NUCLEAIRES

CERTIFICATE OF FINANCIAL SECURITY

FOR THE CARRIAGE

OF NUCLEAR SUBSTANCES

Modèle de Certificat recommandé par le Comité de Direction de l'ENEA
Model Certificate recommended by the ENEA Steering Committee

CERTIFICAT DE GARANTIE FINANCIERE POUR LE TRANSPORT DE SUBSTANCES NUCLEAIRES
CERTIFICATE OF FINANCIAL SECURITY FOR THE CARRIAGE OF NUCLEAR SUBSTANCES

établi conformément à l'article 4(c)   de la Convention de Paris sur la responsabilité civile dans le domaine de l'énergie nucléaire, en date du 29 juillet 1960 et à la loi ....

issued in accordance with article 4(c) of the Paris Convention on Third Party Liability in the Field of Nuclear Energy of 29th July 1960 and the Law ....

I

1. NUMERO DU CERTIFICAT ....
   CERTIFICATE NUMBER

2. NOM ET ADRESSE DE L'EXPLOITANT RESPONSABLE
   NAME AND ADDRESS OF THE OPERATOR LIABLE

   Nom
   Name ....

   Adresse
   Address ....

   ....

3. MONTANT DE LA GARANTIE
   AMOUNT OF THE SECURITY

   ....
   ....

4. TYPE DE GARANTIE
   TYPE OF SECURITY

   ....
   ....
   ....
   ....

5. DUREE DE LA GARANTIE
   DURATION OF THE SECURITY

   ....
   ....

6. DESIGNATION DES SUBSTANCES NUCLEAIRES COUVERTES PAR LA GARANTIE
   NUCLEAR SUBSTANCES IN RESPECT OF WHICH THE SECURITY APPLIES

   ....
   ....
   ....
   ....

7. ITINERAIRE COUVERT PAR LA GARANTIE
   CARRIAGE IN RESPECT OF WHICH THE SECURITY APPLIES

   ....

   ....

8. NOM ET ADRESSE DE L'ASSUREUR (OU DES ASSUREURS) ET (OU) DE LA (OU DES) PERSONNE(S) AYANT
   ACCORDE UNE GARANTIE FINANCIERE
   NAME AND ADDRESS OF THE INSURER (S) AND/ OR GUARANTOR(S)

   Nom
   Name ....

   ....

   Adresse
   Address ....

   ....

   DELIVRE A .... LE .... PAR
   ISSUED IN   *ON*   *FOR AND ON BEHALF OF*

   (a) Le (ou les) garant(s)
        The guarantor(s)

        Designation ....

        Signataire et titre ....
        Signer and title

   (b) L'Etat [le cas échéant]
        The State [where applicable]

        Signataire et titre ....
        Signer and title

II

Je soussigné, certifie que la personne visée au paragraphe 2 est un exploitant au sens de la Convention de Paris.
I hereby certify that the party mentioned in Paragraph 2 is an operator within the meaning of the Paris Convention.

Délivré à .... le .... par
Issued in   *on*   *for and on behalf of*

....

....

....

(L'Autorité publique compétente)
(*The Competent Public Authority*)

L'EXPLOITANT RESPONSABLE
THE OPERATOR LIABLE ....

....

    dont le siège est
    whose address is ....

    ....

    ....

    certifie que le transport de substances nucléaires décrit ci-après est effectué pour son compte et qu'il est visé par la garantie mentionnée dans le Cadre I.
    certifies that the carriage of nuclear substances described hereinafter is carried out on his behalf and that such carriage is covered by the security mentioned in Part I.

DESIGNATION DES SUBSTANCES NUCLEAIRES COUVERTES *PAR* LA GARANTIE
NUCLEAR SUBSTANCES IN RESPECT OF WHICH THE SECURITY APPLIES ....

....

....

....

....

....

....

ITINERAIRE COUVERT PAR LA GARANTIE
CARRIAGE IN RESPECT OF WHICH THE SECURITY APPLIES ....

....

....

....

....

....

Délivré à ....   le ....   par
Issued in   *on*   *for and on behalf of*

....

....

Signature :

L'exploitant responsable
The Operator liable

NOTICE EXPLICATIVE
RELATIVE AU CERTIFICAT DE GARANTIE FINANCIERE POUR LE TRANSPORT DE SUBSTANCES NUCLEAIRES

CADRE I

En-tête

L'en-tête pourra comprendre une référence à l'autorité publique competente du pays qui établit le certificat.

Paragraphe 2

Lorsque, conformément à l'article 4 (d) de la Convention de Paris, la loi nationale prévoit que la responsabilité du transporteur peut être substituée à celle de l'exploitant normalement responsable, et qu'il est fait usage de cette faculté, le nom et l'adresse du transporteur devront remplacer ceux de l'exploitant.

Paragraphe 3

Le montant indiqué pour la garantie doit être par accident; toutefois, s'il n'est pas possible d'obtenir une couverture par accident, il doit être précisé si la couverture est par période ou par voyage. Si le montant global de la garantie résulte de plusieurs garanties différentes, le montant de chacune d'elles doit être précisé. Le montant global de la garantie doit être conforme aux dispositions de l'article 7 (b) et (c) de la Convention. Si la garantie financière résultant d'une assurance ou d'une autre source est insuffisante, les autorités nationales compétentes doivent indiquer l'importance des fonds mis à disposition par l'Etat ou les mesures complémentaires prises par celui-ci.

Paragraphe 4

Le certificat doit mentionner s'il s'agit d'une assurance (et, dans ce cas, préciser le numéro de la police) ou bien s'il s'agit d'une autre forme de garantie financière. Si la garantie est fournie sous plusieurs formes différentes, celles-ci doivent être énumérées y compris, le cas échéant, les fonds publics.

Paragraphe 5

L'inscription «durée de la garantie» doit préciser la date d'effet de la garantie. Il est rappelé, qu'aux termes de l'article 10 (b) de la Convention, l'assureur ou le garant ne peuvent suspendre ou mettre fin à la garantie financière pendant la durée du transport.

Paragraphe 6

La description doit permettre d'identifier de façon précise les substances nucléaires faisant l'objet du transport. Toutefois, dans le cas où l'exploitant est titulaire d'une police d'assurance ou d'autres garanties couvrant en permanence toute une série de transports pendant une période définie, la description donnée au paragraphe 6 pourra être de caractère général, à condition que le Cadre III, d'usage facultatif, soit alors rempli et permette l'identification précise des substances nucléaires faisant l'objet du transport particulier pour lequel le certificat est délivré.

Paragraphe 7

Dans la mesure où les principaux points de passage du transport, en particulier aux frontières, sont connus par avance, ceux-ci doivent être indiqués. Le nom et l'adresse du destinataire pourront éventuellement être précisés.

Paragraphe 8

Lorsque la garantie constituée par une police d'assurance est complétée par une garantie accordée par l'Etat ou un autre garant, leur signature doit figurer au bas du Cadre I.

CADRE II

En certifiant que la personne désignée au paragraphe 2 du Cadre I est un exploitant au sens de la Convention de Paris, les autorités compétentes pourront également faire figurer les garanties fournies par l'Etat, ou les autres mesures prises par lui, pour assurer l'indemnisation des victimes, conformément à la Convention.

CADRE III

Le Cadre III, d'usage facultatif, est rempli par l'exploitant lui-même lorsque la garantie figurant au paragraphe 6 du Cadre I fournit une couverture générale valable pour toute une série de transports. Le Cadre III vise alors le transport particulier dont il donne la description. Le Cadre III ne peut en aucun cas constituer à lui seul un certificat valable et il ne peut être utilisé qu'en complément du Cadre I.

EXPLANATORY NOTICE
ON THE CERTIFICATE OF FINANCIAL SECURITY FOR THE CARRIAGE OF NUCLEAR SUBSTANCES

PART I

Heading

If desired, the heading may include a reference to the competent public authority of the country where the Certificate is issued.

Item 2

Where, in accordance with Article 4 (d) of the Paris Convention, national law provides that the carrier may be liable in place of the operator who would normally be liable and when use is made of that option, the name and address indicated should be that of the carrier rather than that of the operator.

Item 3

The amount of security indicated shall be per incident; if, however, per incident coverage is unobtainable, it must be indicated whether the coverage is per period or per carriage. If the total amount of security has been furnished by more than one source, the amount of each of them should be indicated. The total amount of security must conform to the provisions of Article 7 (b) and (c) of the Convention. If the financial security furnished by insurance or from some other private source is insufficient, the competent national authorities should indicate the funds made available by the State or other supplementary measures taken by the State.

Item 4

The Certificate should stipulate whether the security furnished is by insurance (including in such cases the insurance policy number) or whether such security is furnished in some other form. If security is furnished in several forms, these should be enumerated, including State funds.

Item 5

The entry "duration of the security" must stipulate the date on which such security takes effect. It should be recalled that Article 10 (b) of the Convention provides that no insurer or other financial guarantor shall suspend or cancel the financial security during the period of the carriage in question.

Item 6

The description given of the nuclear substances should be sufficiently complete to enable them to be positively identified. However, where the operator holds an insurance policy or other financial security providing continuous cover for a whole series of carriage for a defined period, a general description may be given in Item 6, provided that Part III, of optional use, is completed and enables the exact identification of the nuclear substances involved in the particular carriage for which the Certificate is delivered.

Item 7

The major points of transit should be indicated where known, notably the crossing of national borders. Where desired, the name and address of the consignee may also be given.

Item 8

Where the State or some other guarantor completes the security furnished by insurance, they must also sign at the bottom of Part I.

PART II

In certifying that the party mentioned in Item 2 of Part I is an operator within the meaning of the Convention, the competent authorities may also include mention of the security furnished by the State or of other measures which it has taken, to ensure the compensation of persons suffering damage, in conformity with the Convention.

PART III

Part III, of optional use, should be completed by the operator himself when the security mentioned in Item 6 of Part I provides general coverage for a whole series of carriage described therein. Part III may, in no case, constitute a valid certificate in itself and is only valid when used in conjunction with Part I.

CONTRACT CLAUSES WITH RESPECT TO
TRANSFRONTIER MOVEMENTS OF HAZARDOUS WASTE

S. Baumgartner, F. Hoffmann La Roche
and P. Tobler, Ciba-Geigy, Suisse

SUMMARY

Transfrontier movements of wastes and waste management operations are a challenge for those responsible for drafting suitable regulations and agreements. In order to propose possible contract clauses we have taken into consideration the contractual practices in Switzerland and the Recommendation and Principles of the OECD. Moreover, it is evident that transport and disposal have to be undertaken in accordance with the laws and regulations applicable in the countries concerned. We have shown that the law on contracts is quite suitable to cover any details left open by the rules and regulations. Being inherently flexible, this law will furthermore allow and facilitate new developments in the management of hazardous wastes.

INTRODUCTION

In trying to pinpoint areas in which particular contract clauses with respect to transfrontier movements of hazardous waste are required, we had to take into account the entire background: i.e., not only the specific regulations in various countries but also the law in general and last, but not least, certain aspects of international civil law. We then described the actual practice in intrafrontier and transfrontier movement in Switzerland. Based on these elements, we tried to specify certain critical areas where properly drafted clauses are of importance to ensure proper handling and disposal of hazardous wastes.

In setting out the problem areas we have used, as an example, mainly Swiss law which we know best. Since Swiss law is essentially similar to other continental civil law, we feel that this approach is permissible and that it will lead to fairly representative results. The legal problems under other continental laws may not quite be the same but they are bound to be fairly similar; and although Common law is quite different, we hope that the presentation will still be useful, as it will allow those versed in Common law to gain some understanding of the specific problems encountered in continental law.

There is another caution to be added. The summary of the Swiss regulations on dangerous wastes is not based on an official draft, since the first official draft is not yet published.

SURVEY ON THE REGULATIONS CONCERNING HAZARDOUS WASTE DISPOSAL
IN THE EUROPEAN COMMUNITY, SWITZERLAND, GERMANY,
FRANCE AND UNITED KINGDOM

1. The European Community

At present two Directives relating to waste disposal are in force: Council Directive of July 15, 1975, on the disposal of waste (75/442/EEC) and Council Directive of March 20, 1978, on toxic and dangerous waste (78/319/EEC). Whereas the Directive of 1975 is concerned with the handling of waste disposal in general, the Directive of 1978 deals with the disposal of hazardous waste within the Community. The Member states are directed to allow the disposal only by disposers who are holders of a licence and whose sites are supervised by the authorities. The producer shall be obliged to dispose of waste only through licensed disposers. Producer, carrier and disposer shall furnish all needed information to the authorities and keep records of this data.

The proposed new Council Directive on the supervision and control of the transfrontier shipment of hazardous waste within the European Community urges further measures which are now in discussion:

-- Notification of the transfrontier shipment to all national authorities concerned;

-- Evidence of safe and appropriate disposal to be approved by the authorities;

-- Consignment notes and, if necessary, use of special roads and transit sites;

-- Special licence for and supervision of the carrier;

-- Special insurance;

-- Absolute liability of the producer until disposal.

The draft Council Directive on the supervision and control of the transfrontier shipment of hazardous waste within the European Community suggests an absolute liability of the producer (Article 15). This article -- and others --are still controversial since this proposed liability seems to be even broader than the concept of strict liability. For a detailed criticism of this proposal, reference is made to the CEFIC report presented at the Seminar.

2. Switzerland

Until recently, there was no specific legislation concerning waste disposal. The disposal itself by dumping was covered by a State permit granted in accordance with the terms of the Federal law on the Protection of the Waters and disposal by other means such as burning or transformation was covered by State law. There were no special transport regulations except the rules on the transport of dangerous goods. On January 1, 1985, the new Federal law on the protection of the environment (7/10/1983) comes into force. A proposed regulation based on this law (Verordnung über den Verkehr mit gefährlichen Abfällen) shall deal with transport and disposal of hazardous waste:

-- Consignment notes and data sheets shall be necessary;

-- Any waste disposer shall need a licence;

-- Import, export and transport shall require a formal procedure;

-- Disposers have to give assurance of acceptance in advance of each shipment from abroad.

This regulation shall be submitted during the summer of 1984 to the States, political parties and interested groups for comments. It is expected that it will come into force at the earliest at the end of the year, but more probably sometime in 1985.

As currently drafted, the regulation covers transfrontier movements of wastes but does not unduly restrict them. In particular, there is no limitation forbidding the export of dangerous wastes if there is a possibility to dispose of them in Switzerland.

3. Germany

The handling of industrial waste is covered by a special law on waste disposal (Gesetz über die Beseitigung von Abfällen vom 5.5.1977). Several additional decrees exist relating to import, shipment and disposal (Abfallbeförderungsverordnung 29.7.1974, Abfalleinfuhr-Verordnung 29.7.1974, Verordnung zur Beseitigung von Abfällen 24.5.1977, etc.). Only licensed disposers with special facilities are allowed to handle hazardous waste. Their declaration of acceptance is required for transporting dangerous wastes. Carriers need a licence for the transport of their wastes and consignment notes with all the relevant information have to be transferred between producer, carrier and disposer.

Export of dangerous wastes is limited to those which cannot be safely disposed of within the country. This interdiction has been known to hinder efficient waste disposal in industrialised border regions.

4. France

Several laws and decrees deal with import, shipment and disposal of hazardous waste (Loi No. 75-633 du 15 juillet 1975 relative à l'élimination des déchets et à la récupération des matériaux, décret No. 77-974 du 19 août 1977 relatif aux information à fournir au sujet des déchets générateurs de nuisances, arrêté du 15 juillet 1983 relatif à l'importation des déchets dangereux et toxiques, arrêté du 15 avril 1945 "règlement pour le transport par chemin de fer, par voies de terre et par voies de navigation intérieure des matières dangereuses)". Any person who produces, imports, carries or disposes of hazardous waste has to furnish all information on the source, quality, quantity, disposal site, etc. to the competent authorities. Producer and disposer are jointly held liable for damage. Dangerous wastes have to be declared to customs and the destination indicated in case of transfrontier shipment (Code des Douanes, art. 95/99). The producer has to furnish the carrier with all relevant information and to provide adequate packing.

5. United Kingdom

The "Control of Pollution Act 1974" provides a coordinated mechanism for waste disposal. It imposes upon county councils the duty to prepare a waste disposal plan for its area, to provide for the disposal licences to ensure that waste treatment and disposal is carried out without unacceptable risk to the environment and to public health. Additionaly, "The Control of Pollution (Special Waste) Regulations 1980" came into force in March 1981. If a producer of hazardous wastes wishes to remove them from the premises where they were produced, he has to prepare a special consignment note, enter thereon a description of the waste and the journey envisaged, and send a copy to the disposal authority for approval and declaration of acceptance. When the waste is collected by the carrier, he completes a declaration on the consignment note to that effect, and the producer declares that he has given any appropriate warning to the carrier. The disposer, upon receipt of the waste, endorses on the notes a certificate confirming that his facility is licensed to dispose of this waste.

CERTAIN ASPECTS OF NATIONAL AND INTERNATIONAL CIVIL LAW

1. General

Although wastes and their handling have been a problem at least since Roman times, remarkably little law has been passed dealing with the specifics of what is termed today as "waste management" in its broadest sense. The elimination of wastes was traditionally held to be a public task, to be performed, if at all, by the authorities (or entrusted, on their behalf, to private parties). Activities creating particularly offensive wastes were banned and allowed only extra muros (as the chemical industry in the 19th century in many places) or in certain parts of the city (e.g., the tanners' quarter in ancient Rome). As to civil law, the main issue always seemed to be the effects the wastes had on neighbours, and there is an extensive body of law on this particular problem, the sources of which are probably lost in the origins of time and which stretches in an almost uninterrupted sequence from the Corpus Juris to the modern codifications. Apart from this particular issue, there is almost nothing to be found in traditional civil law. Therefore, too, the traditional type contracts contained in most codifications do not cover the particularities of waste transport or disposal.

Economically, there is a reason for this lack of legal precedent and even for suitable concept in the law. Concepts, contracts and commercial parties relating to the transport of goods, as an example, were developed to cover the transport of valuable merchandise. Anticipated behaviour of the parties to the contract, indeed their rights and liabilities, were determined based on the fact that the transported goods were worth at least something and would, at the very least, be missed if they disappeared en route. Conversely, the concept of having to pay somebody to take off goods not worth anything and dispose of them in a suitable manner, is almost foreign to traditional law. In short, development of suitable law is a fairly recent matter, since the economic problem has arisen only recently, too. The central question appears to be whether the traditional legal instruments are sufficiently flexible to cope with this new situation or, if not, if the new laws and regulations passed in connection with the transport and disposal of dangerous wastes are sufficient to resolve whatever issues of law remain. Only if these issues have been resolved, does the drafting of suitable contracts become a fairly straightforward matter, since it will be then known what is covered by law and what, in addition, has to be included into the contract.

2. Type and Terms of Contract: the Passing of Property and Title

Civil law, at least under the Swiss Code of Obligations, leaves the parties a great freedom in their choice of terms for their intended contracts. They are not bound by the type contracts set out in the law, such as the purchase contract, contract of work, employment contract, contracts on the transport of goods, etc. Their sole limitations in their choice are that they may not (at the risk of nullity of the particular clause or even the entire contract), infringe the mandatory limits set out in the Code of Obligations or elsewhere (in particular they may not circumvent the regulations covering the handling of dangerous wastes).

It would therefore appear at first sight that contracts concerning the proper management of dangerous wastes should be perfectly feasible and, indeed, a suitable instrument for handling the problems associated with waste management. This is indeed the case, and hardly surprising, since successful waste management has not, with very few regrettable exceptions, awaited the passing of suitable legislation. This is certainly true for Switzerland where the most important facilities for the disposal of dangerous wastes were set up based on civil law. Nevertheless, there are problems and uncertainties. The type contracts provided by the law do not always apply; for instance, the law covering the sale of goods does not allow for a "vendor" who has to pay for inducing the "buyer" to accept the goods. A similar situation prevails with a contract of work supposedly covering the safe disposal by burning or disposing of dangerous wastes. These contracts are, by definition, sui generis, which means, simply put, they they are "something else". This does not define, however, what they are.

There is worse, however, than the problem of suitable drafting or the problem of choosing an appropriate type contract. The new legislation on the handling and the disposal of dangerous wastes has raised the issue of whether the traditional notions of the passing of property, risk and title are still valid in this context. The problem arises from a conflict of basic principles. So far, it has been possible to transfer title, risk and property to a third party. In waste management, this appears to be no longer obvious, at least if the proposed EEC regulation on this topic achieve force of law. Even without this problem, the conflict between concepts remains obvious enough. Civil law provides that risk passes once property and title have passed, as a matter of course. Any exception to the rule will require a basis in the contractual arrangements covering the transaction. Environmental law about to be passed, however, proclaims the opposite. The generator of wastes is supposed to remain responsible and liable for whatever happens after his having disposed, in whatever way, of the dangerous wastes. In fact, as well as probably in law, he is in all material ways no better off as if he had retained risk, property of and title to the wastes, since he is expected to be liable to all damage caused by third parties, to accept the return of the wastes turned over and do whatever the law may require him to do for the ultimate safe disposal of these wastes regardless of the passing of risk, title and property.

Whatever the grounds for this particular legislation covering the management of dangerous wastes may be (and the reasons, if not the means, may be quite reasonable), there appears to exist a conflict of law which makes it difficult to foresee the outcome of any litigation on some of the issues.

3. International Civil Law

The above will be largely sufficient to explain why any case involving international aspects, such as the transborder movement of hazardous wastes, will present particular difficulties under international civil law. Whatever traditional rules may have been developed with regard to type-contracts, it is clear that they would not necessarily apply to the transactions here considered, i.e., the transborder transfer of hazardous wastes. This is particularly true if the contract not only covers the simple transport of dangerous wastes across a border, but also includes the disposal or recycling or the passing on to a third party of such wastes in another country.

Traditionally, international civil law provides certain rules which may be used to determine the law of the contract if there is no choice of law and no established precedent. The characteristic performance is relied on, for instance, in the sale of goods, and it would be tempting to apply the same rule to the waste management contract by claiming that the transporter or disposer is responsible for the characteristic performance. This approach is not entirely convincing since the special duties and obligations to be imposed upon the original producer by the new laws and regulations make it quite clear that the ultimate control of the dangerous wastes and most of what happens with them remain with the producer of these wastes, including the obligation to take or fetch them back if the transport and disposal efforts are not successful. It could be argued, therefore, that the characteristic performance, and indeed the law having the closest connection with the agreement, would be the law of the producer of the wastes, and that therefore this law should apply.

Unfortunately, very little precedent is available for either arrangement available. The proposed new treaties, guidelines, laws and regulations offer no further guidance, either.

THE ACTUAL PRACTICE IN TRANSFRONTIER MOVEMENTS OF HAZARDOUS WASTE: SOME EXAMPLES OF CONTRACTS AND AGREEMENTS

1. Introduction

From the generation of hazardous waste until the final disposal, many actions have to be taken: delivery, collection, transport, acceptance and disposal, the latter comprising, e.g., treatment, incineration, recycling, storing, dumping. Each step and each process can be part of an integrated operation or, conversely, be performed separately, under separate contract, with separate contractors.

In actual practice, we have observed many different arrangements, at least some of which routinely include the transfrontier movements of dangerous wastes. There is no set preference for a particular arrangement. The same company may bring, with own transports, dangerous wastes from a production site abroad and incinerate them on site in Switzerland, while shipping, through third parties, other wastes to a suitable facility abroad, for recycling and participate, for still other types of dangerous wastes, also in the arrangements for permanent disposal with third parties and public authorities.

Waste disposal across the border is a must for the Swiss chemical industry, since not all hazardous wastes can be reasonably disposed of in Switzerland. Conversely, there is, or at least was, considerable import of hazardous wastes from Germany.

Contractual arrangements usually reflect the different ways of handling the problem. Actual arrangements relating to the transfrontier movements of dangerous wastes will sometimes form only a small part of the entire contract. Nevertheless, for practical purposes, the arrangements relating to transport and those relating to the disposal itself can be fairly easily separated. Each part has its particularities and technical aspects which will have to be taken into account. For the part dealing with transport only, it is mainly a question of complying with the specific legislation applicable for the chosen mode of transport, and less of a legal problem. Conversely, the part dealing with the disposal problem requires a great deal of thought and a careful drafting of contracts, as it is not always possible to rely on type contracts.

Another problem which has so far been unresolved, appears to be close to a solution. The international rules concerning the carriage of dangerous goods are being revised to specifically cover also dangerous wastes. It is hoped that this revision will lead to an appropriate inclusion of dangerous wastes into the existing categories, so that their individual properties such as toxicity, flammability, corrosivity, etc., will be relevant for classification. A general chapter covering wastes only would be justified only if dangerous wastes were to present additional dangers in transit due to the mere fact that they are intended for disposal and not for ordinary consumption.

2. The Transport of Dangerous Wastes

As already indicated, the transport part, in its narrowest definition, does not really present all that many unusual legal problems. Normally, the established instruments used by the carriers, the shippers, their agents, etc. are quite suitable for the task at hand. Dealing with the additional paperwork required under the various national and international legislation already passed or about to be passed may create some additional problems but their main impact is to be seen in connection with the practical handling.

A problem to which no definite solution can be given as yet refers to the question of liability for lost goods (in this instance hazardous wastes) and damage thereby caused. The relevant dispositions contained in national laws and international treaties have all been drafted on the assumption that the goods to be transported are of economic value and importance to the party sending or receiving them and address mainly the contingencies "loss or damage to goods". In transporting dangerous wastes, the value of the goods transported is often nil, and the main concern is safe delivery and avoidance of damage to third parties or the environment. It is hard to see, under these circumstances, why there should be limitations of liability of the carrier based on the fictional value of the goods transported or an arbitrary limit, as in the CMR (Convention on the Contract for the International Carriage of Goods by Road).

It is felt that the problem of the liability of the carrier and shipper to third parties should be resolved not for dangerous wastes alone. Rather, the general rules concerning the transport of dangerous goods should be suitably modified and include the transport of hazardous wastes. In this respect, it is not sufficient to simply extend the liability of the producer to cover also these contingencies, since this extension of liability of the producer gives no incentive whatsoever to the carrier or shipper to be more careful in transporting hazardous goods and thereby avoid more accidents.

3. Disposal of Hazardous Waste

As already set out, the transport proper of hazardous wastes across one or more borders usually occurs within the framework of a disposal operation and is often covered by the same contractual arrangements as one part of an overall scheme of waste management. A short review of some arrangements illustrates the diversity of arrangements currently in operation with a single large chemical manufacturer sited in a border area of Switzerland and having manufacturing sites in various neighbouring countries:

-- A contract for the recovery of certain metals from sludge including the transport from the manufacturing site to the refinery abroad in special vessels, by river transport;

-- A contract for the supply of contaminated solvents, to be burnt as substitute fuel in the manufacture of cement;

-- A contract for the disposal at a disposal site abroad (Herfa-Neurode);

-- An arrangement between affiliated companies for the collection of waste sulphuric acid, the conversion thereof into gypsum and the disposal thereof for landfill or the manufacture of cement by a third party;

-- The construction and operation of a disposal site for hazardous wastes as the lead member of a partnership of chemical manufacturers. The day-to-day operation includes the determination of admissible wastes, the operation of the site itself and, for wastes originating within the partnership, general transport arrangements;

-- Acceptance of hazardous wastes from a third party for incineration in special facilities.

In each of the above, at least part of the hazardous wastes involved will cross a border at least once. Choice of the appropriate arrangement for the management of hazardous wastes is determined mainly on technical grounds. In the above cases, the need to cross borders often is a result of the local particularities (closeness of the sites to the border, international operations, diversified operations) and reflects the need to find technically and economically suitable solutions.

Contractual arrangements covering the above situations vary considerably. For some cases, the basic arrangement and the individual transaction are covered by the same contract. In other instances, there is a basic arrangement and the individual transaction is conducted within this framework.

It is perhaps surprising to realise that most of the arrangements work quite satisfactorily with a minimum of formal contracts and instruments. A survey conducted by the authors has revealed that there are, apart from the General Terms of Contract of the professional waste disposers, no uniform contracts, at least not in Switzerland.

PROPOSED RULES FOR DRAFTING CONTRACTS INVOLVING THE
TRANSFRONTIER MOVEMENT OF HAZARDOUS WASTES

1. Choice of Law and Contract

a) Early involvement of a legal practitioner

Experience has shown that the early involvement of legal practitioners in the development of a hazardous wastes management policy is beneficial, especially if such policy includes the transfrontier movement of such wastes. A lawyer will be qualified to point out all the legal risks and thereby provide management with necessary information. In particular, he will be the only person qualified to determine whether innovative schemes for waste disposal can be realised under the law and make management realise that the law in this novel field may be just as flexible and suitable as it is in other areas of commercial law.

For a lawyer dealing with such questions, the early involvement in the drafting and implementing of a suitable policy is equally useful. The legal restrictions will be taken into account in time. The policy to be implemeted will show the proper respect of the law in all areas. At the same time, full use can be made of the flexibility granted by the law.

b) Choice of law

As indicated earlier, there are no established rules in international civil law determining the law applicable to contracts covering the transfrontier movement of hazardous wastes, as soon as these contracts cover more than the simple transport. Conversely, the parties can expect that any reasonable choice of the applicable law, i.e. of a law having a direct relationship to performance under the contract and at least a party thereto, will be upheld.

Therefore, the initial choice of law will be partly a matter of convenience (i.e. choice of law most familiar or most advantageous or both to the choosing party). The choice of law will be partly determined, too, by the type of contracts and terms it permits. It will allow the choosing party to select the law which permits the drafting of the contracts intended, with the least risk of invalidity.

c) Choice of contract

Once the law applicable to the transaction is determined, the choice of contract type and the terms of the contract have to be made. Since there are no type contracts for certain of the transactions considered (at least under Swiss law), the next decision will be whether to assimilate the transaction to an existing contract type or to entirely cover the matter by specific contract clauses. The former technique is generally recommended since it makes the body of law developed for the type contract available for the interpretation of the actual contract in question. The latter technique, though more burdensome, certainly presents the advantage of setting out completely and unequivocally the rights and duties of the parties.

d) Some examples

As already indicated, neither the type contract relating to the sale of goods nor the one relating to work are really applicable to the transactions involving hazardous wastes, although the transactions themselves resemble such contracts. If, therefore, a Swiss company wants to dispose of hazardous wastes abroad (as it may well have to, since there may be no disposal sites available in Switzerland for these wastes), it will probably choose to submit the contract to Swiss law for convenience and assimilate the contract to a contract of sale if the disposer is supposed to recycle or "resell" the hazardous wastes, and to a contract of work if the disposer is supposed to incinerate or dump the hazardous wastes in an approved site. In the first instance, the purpose of the contract is to transfer property, risk and title, to the extent feasible, as in a sale; in the second instance, the purpose of the contract also comprises the performance of some work. The remedies available under a contract of work cover the latter contingency also in that they specifically provide a claim for repairing improperly performed work in addition to any claim for damages, as under the contract for the sale of goods. Conversely, a contract similar to a mandate (Art. 394 of the Swiss Code of Obligations) is to be avoided, if possible, since it requires only best efforts of the agents, and does not give a guarantee of performance. Moreover, it can be terminated, by law, at any time, with immediate effect.

A particular problem concerns the international transport of the hazardous wastes. As already indicated, international conventions, with their limitations on liability, are not really suitable. It is therefore recommended to include transport into the duties of a party and not to make it subject to a separate agreement, thereby ensuring that the party responsible for transport bears the full responsibility also for this segment of its performance.

2. General Drafting Suggestions for Contracts on Transfrontier Movements of Hazardous Wastes

a) A Declaration of Intentions

Although this is not common practice in Switzerland, it is recommended to put a short introduction, setting out the intentions of the parties, at the head of the agreement, as is the practice in Germany and, in a more elaborate form, in the Common-law countries. The introduction will be a valuable help in interpreting the individual clauses of the agreement, and will, to a certain extent, compensate for the lack of precedent in this field.

b) The Technology to be covered

The technical aspects of the disposal of hazardous wastes are largely novel and in a continual flux. There is not as yet a firmly established state of the art, nor is there a uniform terminology in many areas. Moreover, definitions in the relevant national law are by no means identical. It is therefore recommended to make full use of the drafting techniques developed for dealing with such contingencies, i.e. to define, for the purposes of the contract, all the relevant terms and to set out, in all necessary detail, what each part is to do. Often, the latter is achieved by including, in an annex, the disposal concept underlying the particular contract, modified by such additional details as may be necessary to avoid any misunderstandings between the parties.

c) The Influence of National Special Laws on Contracts concerning Hazardous Wastes

In determining performance under the actual contracts, due regard must be given to the special laws on hazardous waste management in any of the countries concerned by the planned transaction. It will be one of the first and foremost tasks in drafting contracts on hazardous waste management to include suitable language incorporating the particular aspects of such national law into the contract. Moreover, general language in the contract should make it crystal clear that neither the letter nor the spirit of the contract may infringe mandatory national law.

d) Disclosure and Secrecy

Most national laws provide for complete disclosure of the relevant information to the competent authorities and most disposers will reflect this requirement in their contracts by requesting that the producer should also disclose all relevant information to them. The latter requirements may cause problems to certain producers of hazardous wastes, however. Wastes from a single source can disclose, provided that they are properly analysed, the entire method of manufacture for a chemical, often a closely guarded trade secret and know-how. The request for divulging such information by a private party, therefore, will be acceptable only if a stringent secrecy obligation is accepted by the party requesting such information. The necessity for such secrecy has been recognised, in the similar field of the notification of new chemicals, by OECD as well as the national authorities. A properly drafted contract will have to resolve this conflict by safeguarding the legitimate interests of both parties involved.

e) Transfer of Title and Liability

It has already been pointed out that transfer of title, property and risk may be influenced by the new legislation on the handling and disposal of hazardous wastes. To some extent, indeed, the traditional concepts of the passing of title, risk, and property will be superseded. It is therefore recommended to cover these aspects in drafting the contract, and also because they largely determine liability under the contract.

f) General Clauses

As already indicated, the traditional, often neglected standard contract clauses have a somewhat different and more meaningful role to assume here. In particular, the points covered before should be taken into consideration and be properly regulated. Among the essentials are choice of law, venue, choice of contract type, etc..

g) Breach of Contract, Termination and Consequences

The necessity to regulate, rather precisely, all matters relating to breach of contract and termination, can hardly be exaggerated. Unlike most contracts covering the sale of goods, there is a particular need to set out in detail these consequences in that the actual fate of the hazardous wastes has to be determined for each contingency. The situation in which the wastes appear to be left derelict while the parties to the contract appear to be squabbling about their rights and duties should never arise, and dispositions setting out the relative rights and duties in arranging for provisional safekeeping of the wastes until their final fate has been determined by the courts having competence, should be included in the contract. Of course, it is impossible to preempt national rules relating to receivership and bankruptcy, but these special situations will have to be considered as "force majeure" which cannot be controlled.

PROPOSED AND ACTUAL CLAUSES EMPLOYED IN TRANSFRONTIER MOVEMENTS OF HAZARDOUS WASTES

In setting out proposed and actual clauses we shall attempt to give guidelines for drafting and not an authoritative wording. First of all, such clauses are unlikely to be drafted in English, since the transborder traffic of hazardous wastes between United Kingdom and Continental Europe is limited. Secondly, even if they are drafted in English, this language will probably be used as the "lingua franca" of modern business and not as the technical legal language for a contract under Common Law, as it properly should. Therefore, the following is restricted to summaries of suggested clauses and certain quotes from actual contracts. It will be the responsibility of the person drafting the actual contracts to determine what will be understood and accepted by the other party and, if need be, by the courts having competence.

1. Declaration of Intentions

A fairly recent example of such a declaration, taken from a contract on the recycling of sulphuric acid, runs as follows:

"X Co. has on its factory premises an accumulation of several thousand tonnes of sulphuric acid waste in a concentration of at least x% H_2SO_4. This sulphuric acid waste is in the opinion of Z. Co. suitable for use in the production of adjuvants. Z. Co. is interested in obtaining sulphuric acid waste from X. Co. The parties therefore agree as follows:"

If the contract is drafted in English, it is recommended to use the technique of the so-called "WHEREAS"-clauses as developed under Common Law. A reservation must be made, however. The declaration of intentions should not cover any stipulation which should properly be contained in the paragraphs of the contract.

2. The Technology to be covered

a) The Definition of Terms

As already set out, there is a need to define, for the purposes of the contract, the terms employed. The contracts we have surveyed contain various solutions, such as the definition of the hazardous wastes by physico-chemical parameters set out in an appendix ("Details of the chemical and physical properties of the waste are set out in the Annex to this contract") to the contract. Here, too, the traditional drafting technique of the Common Law contracts, which provides for an extensive section containing definitions, may be used as an example ("For the purposes of this Agreement, the terms defined shall have the meaning set forth below:").

b) The Definition of Tasks

A clear and complete definition of the technical tasks to be performed may be required if the hazardous wastes are to be treated in a particular way (incinerating, conditioning, reconditioning) or if a certain mode of operating is of essence. The solution to this problem found in a contract consisted of a detailed description of the technical tasks, attached to the contract as an appendix and forming an integral part of the agreement. The agreement itself covered the construction of a site for the disposal of hazardous wastes and the technical answer covered the essential aspects of construction and operation (Vereinbarung vom 11. Juli 1974 zwischen TEUFTAL AG und CIBA-GEIGY AG).

c) The Terms of the Law

Due care has to be exercised in the choice of terms with regard to the applicable law. An established international terminology of laws does not exist as yet. Therefore, it may be necessary to specify, in terms contained in the relevant national law, what is intended by such terms as "handling", "collection", "transport", "incineration", etc. in English, or "prise en charge", "mise en décharge", "dépôt", "enfouissement", "élimination", etc. in French, or "Entsorgung", "Lagerung", "Beseitigung", etc. in German, if these terms or the actions they refer to are covered by the contract.

3. The Main Tasks under the Contract

It is obvious that both parties will want to specify, in reasonable detail, their respective main tasks (and responsibilities). In this respect the drafting of agreements covering the transfrontier movement of hazardous wastes is fairly straightforward in that it consists of an enumeration of tasks and the necessary guarantees for performance. Here, too, the instructions for dealing with the reporting requirements imposed by the various national laws have to be included, as well as the rules and regulations on transport and packaging, to name but a few.

Although it might be interesting to look at the details of such contracts, a comparison has shown that they are fairly specific to the actual problem and the parties they refer to and that they are not generally usable as examples. Therefore, no sample clauses are here contained.

4. Disclosure and Secrecy

Most contracts surveyed insist on some form of disclosure by the producer of the hazardous wastes. The following clauses are fairly typical:

-- "The firm guarantees the properties of the waste in accordance with the details given on Form A (...). Notwithstanding the provisions of the law, the firm shall be responsible for all damage resulting from a) incorrect or incomplete information in Form A, b) failure of the waste or its properties to comply with the description given in Form A, c) failure of the packaging of the waste to comply with Form A or with the agreed provisions in the written acknowledgement of receipt" (General business conditions of Untertage-Deponie Herfa-Neurode der Kali und Salz AG).

-- "The producer guarantees that the waste delivered complies in type, composition and hazard with the information given on the corresponding "declaration of accountability" and the relevant consignment note and the sample provided" (General business conditions for the disposal of special waste in Hessen).

-- "The producer shall prior to delivery notify the contractor in writing of the exact nature of the waste material involved and draw his attention to any hazard associated with the material." (General business condition H 76, B H 76, Süd-Müll GmbH & Co. KG, Frankenthal)

-- "If the Contractor (disposer) shall discover that any information as to the nature and chemical composition of the waste given by the producer differs from the actual nature and chemical composition of the waste to a material degree, the Contractor shall promptly notify the producer of his findings." (Agreement Roche Products Ltd/Modern Disposals Ltd)

Some contracts we are aware of specify, in addition to the above, the provision of samples or trial quantities for testing or trial operations (e.g., a contract on the use of contaminated solvents as a substitute fuel in the manufacture of cement).

In the cases where the nature and composition of wastes might divulge commercial and trade secrets, the following secrecy obligation on the party receiving the wastes might be useful:

"X undertakes to keep secret and therefore not to disclose to any third party (except reliable employee under secrecy obligation) and to make no commercial use of any information it has received from Y without the express written consent of the disclosing party. This undertaking does not apply if, and to the extent that the receiving part is able to prove that:

a. At the time of communication it already had the information; or

b. The information has become generally available through publication or otherwise, through no violation of this agreement; or

c. After the date of this undertaking, it has received the information from a third party whose direct or indirect source is not the disclosing party.

Notwithstanding anything herein contained, the receiving party may divulge all information received to the competent authorities if this is mandated by law."

5. Transfer of Title, Liabilities

a) Transfer of Title, Risk and Property

Very few agreements clearly specify the exact moment at which title, risk and property pass. The following may be quoted:

-- "(...) Property and risk in the waste shall pass to the disposer at the point of each collection by the disposer."

-- "The stored waste shall -- in the absence of contrary written agreement -- become the property of K + S (disposer) three years after delivery." (General business conditions of Untertage-Deponie Herfa-Neurode der Kali und Salz AG.)

As already indicated, it is recommended to be quite specific on this issue. If a clause of the above type is not desired, the INCO terms of the International Chamber of Commerce could provide an alternative.

b) Liability

The issue of liability is often the subject of heated discussions amongst lawyers. Amongst the clauses found in general Terms of Trade, the following may be quoted:

-- "The disposer shall maintain in force with an insurance company during the currency of this agreement a policy of insurance in respect of Public Liability for such sum and limits of indemnity as may be agreed from time to time (...) but not less than (...). Such insurance shall be extended to indemnify the producer against any claim for which the disposer may be legally liable (...)."

-- "K + S (disposer) shall be liable for any wrongful act, up to the amount of cover provided in its liability insurance. To the same extent K + S shall indemnify the firm (producer) against all claims by third parties based on wrongful acts of K + S in the performance of this contract (or its agents)." (General business conditions of Kali + Salz AG).

-- "The producer shall be responsible for damage resulting from the use of unsuitable or defective containers." (General business conditions special waste Hessen).

Other contracts contain exclusions of liability of the producer, except for specific contingencies, such as:

"X guarantees that the delivered sulphuric acid waste:

a) will have a minimum content of 70 per cent H_2SO_4;
b) does not contain any solvents, and
c) does not derive from the manufacture of herbicides or insecticides.

No further guarantee is given."

Conversely, the contract on the operation of the disposal site for hazardous wastes in KOELLIKEN names as one of the purposes of the partnership the assumption of all liability under public and private law resulting from the operations of the disposal site:

"The Consortium accepts:

a) ....

b) Criminal (Article 6 GSchG) and civil liability (Article 6 GSchG) and civil liability (Article 36 GSchG, Article 58 OR, Article 679 ZGB), provided the damage is not caused by undertakings responsible for construction and operation or by the acts of third parties."

(Art. 1 of the above Partnership Agreement between the Cantons Aargau and Zurich, the City of Zurich and CIBA-GEIGY Limited, the latter also acting on behalf of the other chemical manufacturers in Basle, relating to the disposal site at KOELLIKEN.)

As in other fields, the thorny issue of allocating liabilities can often be defused by providing for proper insurance cover. Full use should be made of the facilities and the cover offered by underwriters.

6. General Clauses

a) Applicable Law

The traditional clauses will be quite adequate: there is no need to repeat them here. The only issue which has to be considered is a reference to mandatory national law affecting performance hereunder, if that particular law is not chosen as the law of the contract.

b) The Choice of Contract

As pointed out earlier, there may be a need to determine also the type of contract to which the transaction should be assimilated. The actual text may be quite simple, as an example shows ("Elsewhere the rule of the works contract shall apply") contained in the never implemented Draft Terms for the Incineration of Hazardous Wastes, CIBA-GEIGY Limited, 1982.

c) Venue and Arbitration

The choice of venue will be largely at the convenience of the parties. It will be important only if preliminary measures destined to safeguard the hazardous wastes have to be requested before the court having competence. For this reason, too, it is doubtful whether arbitration is the proper way of resolving disputes relating to the fate of hazardous wastes. In any case, the recourse to the courts for preliminary measures safeguarding the hazardous wastes must be reserved, if arbitration is chosen.

d) Receivership and Bankruptcy

For this contingency, a similar right for preliminary measures should be instituted, to the limited extent this is possible under the appropriate law. Once the official receiver or the bankruptcy administration works properly, there should be no further problems since they act in an official function.

e) Validity of Individual Clauses

A clause should be inserted covering, in the usual manner, the consequences of one or more clauses of the agreement being void; after all, there is a certain likelihood of this occurring. The standard version of this clause can be completed by the addition that it will be the duty of the court having competence to determine which of the surrogate clauses possible shall apply to the contract if the parties themselves are unable to agree.

f) Whole Agreement Clause

This standard clause, too, may have its special significance here since it serves to demonstrate clearly what is part of the contract and, more importantly, what is not part of the agreed transaction. It thereby facilitates rigorous policing of performance under the agreement.

g) Force Majeure and Hardship

The contingency of force majeure should be covered in a manner appropriate to the transaction at hand. It should cover, in particular, the contingency of new restrictions and rules being passed.

Some of the contracts we have examined contain a clause which is essentially a hardship case. This clause provides that the disposer may return, without breach of contract, the hazardous wastes accepted for disposal if the disposal activities turn out to be technically or economically more onerous than foreseen. It is highly recommended to determine parameters for this contingency.

7. Breach of Contract and Termination

a) Breach of Contract

Contracts relating to the transfrontier movements of hazardous wastes should be drafted in a manner that breaching them would be difficult and extremely unrewarding. The former is achieved by clear and complete drafting and the institution of suitable control mechanisms and rights; to cover the consequences of a breach it will be important not to accept too many limitations of liability, such as the ones provided for international carriage of goods earlier referred to. Under Swiss law, there is hardly a need for additional liability provisions in the law; all that is needed, really, is proper use of and reliance on the law to create full liability.

A section covering breach of contract should specifically regulate the fate of the hazardous wastes covered by the contract. Its actual scope will be determined according to the particularities of the transaction. The clause is generally in the interest of both parties.

b) Termination

Termination, and the consequences thereof, should be covered in a similar manner. When the contract runs out or is terminated, each party should be quite clear as to what, if any, its remaining obligations are. This would become more and more important if the current efforts to extend the liability of the producer should be successful.

8. Permits and Compliance

An area often overlooked relates to the allocation of responsibility for obtaining the necessary permits and authorisations. For transfrontier movements of wastes, it is often useful to include a clause setting out the duties of the parties. If there is a list of tasks, then the contents of this section can be added to this list.

9. Summary

It had been hoped to establish, in this section, a clear view of a firm contractual practice relating to transfrontier movements of hazardous wastes. Actually, the situation is not quite as simple as that. The bulk of hazardous wastes is moved across borders based on transport agreements quite similar to those used for goods. This situation may well change and the contract then necessary will have to cover most or all of the contingencies outlined in this section. Moreover, as our waste management becomes more complex and more successful in achieving its aim, the simple transport of hazardous wastes will more and more become a small part of an integrated waste management operation. The challenge for those responsible for drafting suitable agreements will be to cover by agreement those integrated waste management operations and not only the transfrontier movements of hazardous wastes which is but a part of these operations. We hope that the few drafting hints will be of use.

CONCLUSIONS

Proposing possible contract clauses, we have taken into consideration, as background, the Decision and Recommendation of the OECD Council on Transfrontier Movements of Hazardous Wastes as adopted on 1st February 1984. Furthermore, transport and disposal have to be undertaken in accordance with the laws and regulations applicable in the countries concerned.

The contracting parties (generator, disposer) will have to obtain the necessary authorisations to perform their activities and assume responsibility for the proper waste management. The clauses discussed assume the existence of such national legislation, although they have been partly developed before the coming into force of such law.

Finally, we would like to point out that the carriage of hazardous wastes is not fundamentally different from the carriage of other hazardous substances. Therefore, packaging, labelling, transport, insurance and transfrontier movements should be essentially regulated in the same way as the corresponding movements of hazardous goods (e.g., RID/ADR) with as few additions to the regulations as reasonably defensible. Furthermore, the regulations should be clear and as simple as possible, to render them easily applicable and to facilitate control.

We hope to have shown that the law on contracts is quite suitable to cover any details left open by such regulations. Being inherently flexible, it will furthermore allow and even facilitate new developments in waste management.

OECD SALES AGENTS
DÉPOSITAIRES DES PUBLICATIONS DE L'OCDE

ARGENTINA – ARGENTINE
Carlos Hirsch S.R.L., Florida 165, 4° Piso (Galería Guemes)
1333 BUENOS AIRES, Tel. 33.1787.2391 y 30.7122

AUSTRALIA – AUSTRALIE
Australia and New Zealand Book Company Pty, Ltd.,
10 Aquatic Drive, Frenchs Forest, N.S.W. 2086
P.O. Box 459, BROOKVALE, N.S.W. 2100. Tel. (02) 452.44.11

AUSTRIA – AUTRICHE
OECD Publications and Information Center
4 Simrockstrasse 5300 Bonn (Germany). Tel. (0228) 21.60.45
Local Agent/Agent local :
Gerold and Co., Graben 31, WIEN 1. Tel. 52.22.35

BELGIUM – BELGIQUE
Jean De Lannoy, Service Publications OCDE
avenue du Roi 202, B-1060 BRUXELLES. Tel. 02/538.51.69

CANADA
Renouf Publishing Company Limited,
Central Distribution Centre,
61 Sparks Street (Mall),
P.O.B. 1008 - Station B,
OTTAWA, Ont. KIP 5R1.
Tel. (613)238.8985-6
Toll Free: 1-800.267.4164
Librairie Renouf Limitée
980 rue Notre-Dame,
Lachine, P.Q. H8S 2B9,
Tel. (514) 634-7088.

DENMARK – DANEMARK
Munksgaard Export and Subscription Service
35, Nørre Søgade
DK 1370 KØBENHAVN K. Tel. +45.1.12.85.70

FINLAND – FINLANDE
Akateeminen Kirjakauppa
Keskuskatu 1, 00100 HELSINKI 10. Tel. 65.11.22

FRANCE
Bureau des Publications de l'OCDE,
2 rue André-Pascal, 75775 PARIS CEDEX 16. Tel. (1) 524.81.67
Principal correspondant :
13602 AIX-EN-PROVENCE : Librairie de l'Université.
Tel. 26.18.08

GERMANY – ALLEMAGNE
OECD Publications and Information Center
4 Simrockstrasse 5300 BONN Tel. (0228) 21.60.45

GREECE – GRÈCE
Librairie Kauffmann, 28 rue du Stade,
ATHÈNES 132. Tel. 322.21.60

HONG-KONG
Government Information Services,
Publications/Sales Section, Baskerville House,
2nd Floor, 22 Ice House Street

ICELAND – ISLANDE
Snaebjörn Jönsson and Co., h.f.,
Hafnarstraeti 4 and 9, P.O.B. 1131, REYKJAVIK.
Tel. 13133/14281/11936

INDIA – INDE
Oxford Book and Stationery Co. :
NEW DELHI-1, Scindia House. Tel. 45896
CALCUTTA 700016, 17 Park Street. Tel. 240832

INDONESIA – INDONÉSIE
PDIN-LIPI, P.O. Box 3065/JKT., JAKARTA, Tel. 583467

IRELAND – IRLANDE
TDC Publishers – Library Suppliers
12 North Frederick Street, DUBLIN 1 Tel. 744835-749677

ITALY – ITALIE
Libreria Commissionaria Sansoni :
Via Lamarmora 45, 50121 FIRENZE. Tel. 579751/584468
Via Bartolini 29, 20155 MILANO. Tel. 365083
Sub-depositari :
Ugo Tassi
Via A. Farnese 28, 00192 ROMA. Tel. 310590
Editrice e Libreria Herder,
Piazza Montecitorio 120, 00186 ROMA. Tel. 6794628
Costantino Ercolano, Via Generale Orsini 46, 80132 NAPOLI. Tel. 405210
Libreria Hoepli, Via Hoepli 5, 20121 MILANO. Tel. 865446
Libreria Scientifica, Dott. Lucio de Biasio "Aeiou"
Via Meravigli 16, 20123 MILANO Tel. 807679
Libreria Zanichelli
Piazza Galvani 1/A, 40124 Bologna Tel. 237389
Libreria Lattes, Via Garibaldi 3, 10122 TORINO. Tel. 519274
La diffusione delle edizioni OCSE è inoltre assicurata dalle migliori librerie nelle
città più importanti.

JAPAN – JAPON
OECD Publications and Information Center,
Landic Akasaka Bldg., 2-3-4 Akasaka,
Minato-ku, TOKYO 107 Tel. 586.2016

KOREA – CORÉE
Pan Korea Book Corporation,
P.O. Box n° 101 Kwangwhamun, SÉOUL. Tel. 72.7369

LEBANON – LIBAN
Documenta Scientifica/Redico,
Edison Building, Bliss Street, P.O. Box 5641, BEIRUT.
Tel. 354429 – 344425

MALAYSIA – MALAISIE
University of Malaya Co-operative Bookshop Ltd.
P.O. Box 1127, Jalan Pantai Baru
KUALA LUMPUR. Tel. 577701/577072

THE NETHERLANDS – PAYS-BAS
Staatsuitgeverij, Verzendboekhandel,
Chr. Plantijnstraat 1 Postbus 20014
2500 EA S-GRAVENHAGE. Tel. nr. 070.789911
Voor bestellingen: Tel. 070.789208

NEW ZEALAND – NOUVELLE-ZÉLANDE
Publications Section,
Government Printing Office Bookshops:
AUCKLAND: Retail Bookshop: 25 Rutland Street,
Mail Orders: 85 Beach Road, Private Bag C.P.O.
HAMILTON: Retail: Ward Street,
Mail Orders, P.O. Box 857
WELLINGTON: Retail: Mulgrave Street (Head Office),
Cubacade World Trade Centre
Mail Orders: Private Bag
CHRISTCHURCH: Retail: 159 Hereford Street,
Mail Orders: Private Bag
DUNEDIN: Retail: Princes Street
Mail Order: P.O. Box 1104

NORWAY – NORVÈGE
J.G. TANUM A/S
P.O. Box 1177 Sentrum OSLO 1. Tel. (02) 80.12.60

PAKISTAN
Mirza Book Agency, 65 Shahrah Quaid-E-Azam, LAHORE 3.
Tel. 66839

PORTUGAL
Livraria Portugal, Rua do Carmo 70-74,
1117 LISBOA CODEX. Tel. 360582/3

SINGAPORE – SINGAPOUR
Information Publications Pte Ltd,
Pei-Fu Industrial Building,
24 New Industrial Road N° 02-06
SINGAPORE 1953, Tel. 2831786, 2831798

SPAIN – ESPAGNE
Mundi-Prensa Libros, S.A.
Castelló 37, Apartado 1223, MADRID-1. Tel. 275.46.55
Libreria Bosch, Ronda Universidad 11, BARCELONA 7.
Tel. 317.53.08, 317.53.58

SWEDEN – SUÈDE
AB CE Fritzes Kungl Hovbokhandel,
Box 16 356, S 103 27 STH, Regeringsgatan 12,
DS STOCKHOLM. Tel. 08/23.89.00
Subscription Agency/Abonnements:
Wennergren-Williams AB,
Box 13004, S104 25 STOCKHOLM.
Tel. 08/54.12.00

SWITZERLAND – SUISSE
OECD Publications and Information Center
4 Simrockstrasse 5300 BONN (Germany). Tel. (0228) 21.60.45
Local Agents/Agents locaux
Librairie Payot, 6 rue Grenus, 1211 GENÈVE 11. Tel. 022.31.89.50

TAIWAN – FORMOSE
Good Faith Worldwide Int'l Co., Ltd.
9th floor, No. 118, Sec. 2,
Chung Hsiao E. Road
TAIPEI. Tel. 391.7396/391.7397

THAILAND – THAILANDE
Suksit Siam Co., Ltd., 1715 Rama IV Rd,
Samyan, BANGKOK 5. Tel. 2511630

TURKEY – TURQUIE
Kültur Yayinlari Is-Türk Ltd. Sti.
Atatürk Bulvari No : 191/Kat. 21
Kavaklidere/ANKARA. Tel. 17 02 66
Dolmabahce Cad. No : 29
BESIKTAS/ISTANBUL. Tel. 60 71 88

UNITED KINGDOM – ROYAUME-UNI
H.M. Stationery Office,
P.O.B. 276, LONDON SW8 5DT.
(postal orders only)
Telephone orders: (01) 622.3316, or
49 High Holborn, LONDON WC1V 6 HB (personal callers)
Branches at: EDINBURGH, BIRMINGHAM, BRISTOL,
MANCHESTER, BELFAST.

UNITED STATES OF AMERICA – ÉTATS-UNIS
OECD Publications and Information Center, Suite 1207,
1750 Pennsylvania Ave., N.W. WASHINGTON, D.C.20006 – 4582
Tel. (202) 724.1857

VENEZUELA
Libreria del Este, Avda. F. Miranda 52, Edificio Galipan,
CARACAS 106. Tel. 32.23.01/33.26.04/31.58.38

YUGOSLAVIA – YOUGOSLAVIE
Jugoslovenska Knjiga, Knez Mihajlova 2, P.O.B. 36, BEOGRAD.
Tel. 621.992

Les commandes provenant de pays où l'OCDE n'a pas encore désigné de dépositaire peuvent être adressées à :
OCDE, Bureau des Publications, 2, rue André-Pascal, 75775 PARIS CEDEX 16.

Orders and inquiries from countries where sales agents have not yet been appointed may be sent to:
OECD, Publications Office, 2, rue André-Pascal, 75775 PARIS CEDEX 16.

68236-12-1984